December 1992

A Handbook of Public Speaking for Scientists and Engineers

A Handbook of Public Speaking for Scientists and Engineers

Peter Kenny

Institute of Physics Publishing
Bristol and Philadelphia

British Library Cataloguing in Publication Data
Kenny, Peter
    A handbook of public speaking for scientists
    and engineers
    1. Public speaking
    I. Title
    808.5′1   PN4121

    ISBN 0-85274-553-2

First published 1982
Reprinted 1983, 1984, 1985, 1992

Published by IOP Publishing Ltd, a company wholly owned
by the Institute of Physics, London.
Techno House, Redcliffe Way, Bristol BS1 6NX, England

US Editorial Office: The Public Ledger Building, Suite 1035,
Independence Square, Philadelphia, PA 19106, USA

Text set in 11/13 pt Linotron 202 Plantin
First printed in Great Britain at The Pitman Press, Bath
Reprinted in Great Britain by J W Arrowsmith Ltd, Bristol

To
My wife Joan
and
Our two sons Stephen and Richard

'The time has come,' the Walrus said,
'To talk of many things:
Of shoes—and ships—and sealing wax—
Of cabbages and kings—
Of why the sea is boiling hot—
And whether pigs have wings.'

Lewis Caroll, *Through the Looking Glass*

Contents

Part II
Ways of Improving

Part III
Particular Events

Epilogue

Appendices

1 Introduction

The story of the Emperor's new clothes is a very effective way of illustrating the fact that people observe what they think they are expected to observe. The circumstances of the story are somewhat far-fetched, however, and as a consequence we rarely identify ourselves with those foolish spectators. As scientists and engineers we are too wise to be taken in so easily. After all, the scientific method is based on careful observation, and we are constantly aware of the distinction between subjective and objective evaluations.

However, there is one area of activity in the world of science and technology in which we do in fact behave rather like the Emperor's subjects. That is the area of public speaking. Most of us have attended conferences, seminars or symposia to listen to other scientists describing their investigations or bringing us up to date in a particular field of interest. We have all applauded speakers, even complimented them on their presentations. We have of course distinguished between good speakers and poor speakers, but how good were the best? And what was the average standard like? Could you understand what the speakers were getting at, or did you have to rely on reading the preprint? Perhaps the speaker read the preprint aloud for you, and gradually speeded up as he ran out of time. Did anyone keep to time? Was the discussion period encroached upon so that there was time for only two questions? The feelings of frustration were no doubt relieved by several humorous incidents. The odd slide would be shown upside down. The pointer would be used not only as a pointer but as a baton to wave at the audience and as a crutch to support a tiring speaker. Inevitably someone would unconsciously attempt to climb up the pointer as if to demonstrate a variant of the Indian rope trick.

You may think that I am exaggerating. If you believe that the standard of public speaking among scientists and engineers is satisfactory and that you can achieve the required standard when it is your turn to speak, then this book is not for you. My reason for writing this book is my belief that the majority of scientists and engineers consider the standard of public speaking within their professions to be low. The low standard is tolerated because few believe that they can do any better themselves and because the level of expectation is low. If everyone else says the Emperor made a fine presentation of his paper, who is going to argue, particularly if the Emperor is a recognised expert in his subject? It is not surprising therefore that there is little documentary evidence as to the existence of a problem. I had come to the conclusion that I would be unable to quote a reference in support of my case, but at a late stage in the preparation of this book I was sent a copy of a short article from *Electronic Components*.[1] The article appeared, anonymously, in 1969 and the author suggested that the situation was serious enough to question the value of continuing to use technical conferences as a means of communication between technologists.

At college or university we soon discovered that our lecturers were not necessarily good lecturers. After all, they were not employed primarily to lecture; they were employed on the basis of their standing as scientists. Part of the challenge to students was to achieve a standard of knowledge and understanding in spite of some poor lectures. Such a system no doubt has advantages, but it produces generations of young scientists and engineers who have no expectation of listening to many competent speakers on technical subjects.

The great irony is that we *are* in fact taught the importance of good communication in technical matters, but the teaching is restricted to the written word. Our skill in describing an experiment is tested at an early age. We are taught to use a logical format: apparatus, method, procedure, results, discussion and conclusions. In our employment we produce reports which are carefully vetted by our superiors for clarity, brevity, logic, lack of ambiguity, etc. We prepare papers for publication, and editors and referees examine them word by word to maintain high standards. But when we transfer from written to verbal communication the situation is quite different. No one examines what we are going to say at a conference or how we are going to say it, and no one criticises what we have said.

Sometimes at conferences the discussion is published. The editor sends the text of the contribution to the questioner. He transfers what he said into a written contribution. This is then what he should have said or what he wishes he had said. The author on receipt of this amended question then writes a reply. Anyone who has been involved in this procedure will readily confirm that there is often a very tenuous link between what was said at the conference and what is recorded in the proceedings.

It was this anomaly between attitudes to written and verbal communication that lead me to an interest in public speaking. I was concerned about the illogicalities in the system. A technical document to be tabled at a meeting was subjected to a rigorous examination and it was taken for granted that changes were likely to be required before a final version was produced. On the other hand a verbal presentation was left very much to chance. Perhaps it was understood and accepted, misunderstood and not accepted, or even misunderstood and accepted. Not only was I irritated by such a system but I was equally irritated by the knowledge that I could not communicate as effectively when speaking to members of an audience as I could when writing a document for their examination.

I soon discovered that with a modest degree of study it is possible to improve one's public speaking ability. Indeed the improvement is beyond expectation. The purpose of this book is to pass on to others the knowledge which, as I have found from personal experience, can bring about great improvements. I have not become an expert public speaker: indeed it could be argued that great orators are born and would probably have difficulty in explaining to others how to speak well in public. The main qualification I have to justify the writing of this book is that I know what the problems are and I know how they can be overcome. I have suffered fears and humiliations concerned with public speaking and the memories have not faded. Those experiences have gone never to return.

I have highlighted at the outset the poor standard of public speaking among scientists and engineers for a particular reason. Because the standard is poor, even a slight improvement is noticeable. It is not difficult to become better than average: in the land of the blind the one-eyed man is king. To become significantly better than average is I suspect within the ability of anyone. Whether one could become a great speaker by study I rather doubt. The law of diminishing returns sets in and most students of the subject reach a plateau. That plateau represents, nevertheless, considerable personal improvement and is well above the average level exhibited by those who have not studied the subject.

This book is based very much on my own experiences as a scientist, at first struggling with the problems of public speaking, then learning techniques under various teachers, and finally applying those techniques in my professional and social activities. In addition I have drawn extensively on class material that I have put to the test in teaching public speaking to groups of adults of mixed backgrounds and interests. I would like to record my thanks to my students who have studied, practised and put to the test the techniques that I have developed. Thanks are due also to the many colleagues and fellow scientists who have, often unknowingly, provided me with examples of speaking technique. To the various teachers I have had I express my gratitude, especially to J J Scarratt ('Jos'), a first-class teacher of public speaking, who first showed me the way to improvement. I am also most grateful to Professor P L Kirby who not only made constructive comments on my manuscript but encouraged me in my belief that there is room for improvement in standards of presentation at technical gatherings.

If you are an accomplished speaker then I am afraid this book is not for you. If you think you are an accomplished speaker then again it is not for you. If you consider that your skill in public speaking is inadequate and you would like to do better next time, then read on!

2 Scope and Plan of Book

This book is a handbook rather than a textbook in the sense that it is written for people who are busy. It can be used for a talk that has to be given tomorrow or in six months time, or as a means of bringing about increases in skill over a longer period. A wide range of ability in public speaking is represented among scientists and engineers and therefore among potential readers. The contents are designed to cater for those lacking any knowledge or experience, but at the same time are organised so that the more experienced can readily extract helpful hints and techniques.

It is perhaps necessary to define public speaking at the outset. So far as this book is concerned I have adopted the view that public speaking is one-way verbal communication from a speaker to a number of other people, in a situation that demands a degree of formality. It is one-way in that the speaker is not usually interrupted. In this it differs from conversation, as it does in requiring a degree of formality. The degree of formality may be very slight, but nevertheless very real as our nerves will readily testify. Public speaking differs from acting in that the speaker is communicating his own message. The number of listeners could be as few as two though most people would, perhaps mistakenly, reduce such a situation to one of conversation.

It should be explained that public speaking is not primarily

concerned with pronunciation or accent, or English grammar.
Such things cannot be ignored because verbal communication
of any kind must be intelligible. Having assumed, however,
that the speaker can at least make sounds that are meaningful
to the listeners, public speaking concerns itself with the most
effective ways of communicating ideas to several people at
once.

The book is divided into three parts. Part I covers the
tactics; that is to say it deals with the preparation and
presentation of a speech. The treatment is generally applicable, but in order to cover all requirements it has been
necessary to have in mind conferences with fairly large
audiences. Many of the points covered will not apply to
smaller or less formal gatherings. Part II deals with strategy. It
is for those who wish not only to cope with a specific event but
to improve over a period of time. Improvement comes from
exercises and practice and inevitably needs time to be devoted
to it. Nevertheless, even for those with little time or ambition,
some of the chapters, particularly Chapter 8 on confidence,
are worth a quick read through. In Part III guidance is given
in dealing with particular kinds of events. Although there is
some repetition in Part III, it will generally be necessary to
study Part I in conjunction with the appropriate chapter in
Part III.

The book concludes with five appendices. Appendix 1 lists
the references cited, some with comments, that may be useful
to the reader for further study. Appendix 2 gives in flow-chart
form a summary of the steps involved in preparation of a
speech from its first conception to its moment of delivery. The
flow-chart layout allows a rapid check to be made to ensure
that nothing has been overlooked. Appendix 3 consists of a
checklist designed for speech analysis. The use of this checklist is described in Chapter 15. Appendix 4 gives details of
examinations in public speaking that can be entered for at
various centres. The value of such examinations is discussed in
Chapter 16. Appendix 5 consists of lists of quotations classified under various branches of science. Each quotation has a

link with science but is taken from a non-technical and perhaps surprising source. The lists are provided to illustrate the kind of material speakers should collect for use in preparing interesting talks. Chapter 11 discusses the use of such material in detail.

Part I

Preparation and Presentation

3 The Problem

3.1 THE SPEAKER

One of the popular newspapers reported some years ago the results of a survey that had been carried out[2] in which people had been asked to state the things they feared most. Rather than spiders, darkness or heights heading the list, it was found that the most common fear was that of speaking in front of an audience.

What is the origin of this fear? The main cause is lack of confidence on the part of the speaker. The question of confidence is dealt with fully in Chapter 8 but suffice it for now to say that proper preparation and presentation is the first step towards establishing confidence. The second cause of fear is the lack of practice. In fact public speaking is one of those activities that cannot be carried out in a practice situation: it is always either the real thing or not realistic. The third cause of the fear of speaking in public is the lack of proper feedback. No one, not even your best friend, will tell you honestly how you performed. You are forced to live forever with your own uncertainties and subjective evaluation, an evaluation more-over which is made at a time of strain.

Because the fear of speaking in public is so widespread, there is a general acceptance of low standards. Who indeed will throw the first stone? In an atmosphere of acceptance of low standards there is little encouragement given to those who could improve. So generation of speakers follows generation, each carrying its burden of fear and making feeble attempts to 'laugh-off' any suggestion of a serious problem that needs treatment.

3.2   THE MATERIAL

Scientists and engineers are taught to communicate by writing. The technical report has its own format and its own language. The use of the passive voice was traditionally taught though there are signs of change here. The consequence is that the ideas themselves that need to be communicated become embodied, visualised and expressable only in such format and language. I am not speaking here of scientific terminology. It is perfectly proper to define new expressions (standard deviation, microprocessor, etc) or to attribute special scientific meaning to particular words (significance, element, etc). It is the style of scientific writing that I am referring to, for example: 'in order to prevent evaporation, care was taken to ensure that the vessel was tightly stoppered'.

Because of this deeply rooted training there is a natural tendency to prepare material for verbal presentation in a similar style. The speaker is then attempting to speak in a way that is completely foreign to him, and his listeners are having to embark on a mental process of simultaneous translation. The problem is least troublesome when the specialist addresses other specialists. When the specialist addresses a non-scientific audience the problem is most apparent. The speaker will of course have taken care to exclude or define technical jargon but he will generally fail to appreciate that the style and format he is attempting to use is completely unfamiliar to his audience. It is often said that scientists have too little influence on Society. One wonders if this is related to this communication problem.

3.3   THE AUDIENCE

An audience may be captive in a physical sense but it is rarely totally captive mentally. It is safe to assume that only a small proportion of an audience will make great efforts to understand a speaker. The majority are perhaps prepared to listen for a while, but if meaningful messages do not enter the brain then the attention will wander. A few are perhaps resigned to the fact that they will gain nothing from what is said and one or two are probably present only by mistake or circumstance.

It is probably not so generally recognised that even those members of the audience who are dedicated to listening to the speaker suffer from the lack of attention. The so-called attention curve, determined from studies in classroom situations is produced in figure 1.[3] Although nearly 100% of a class may be giving full attention after five minutes, the curve falls steeply so that only 20% may be giving full attention after thirty minutes. Experiments designed to measure the amount of recall that members of an audience are capable of give results no more comforting to the speaker. After a fifteen minute lecture the average immediate recall can be 40% and after a forty-five minute lecture as little as 20%.

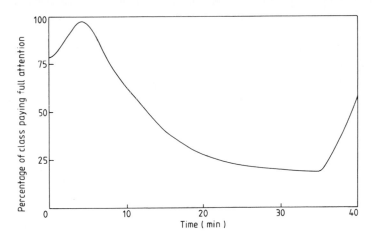

Figure 1   The attention curve.[3]

Notice how different the situation is with regard to written material. The audience, which now consists of the readers of the document, will still suffer from wandering attention and lack of interest, but the reader can go back a few sentences, speed up or slow down as required. He need not even read the document in the sequence in which it was written. In fact he probably rarely does. A more common pattern of absorbing a written document would be a quick flick through to get the feel of the length and complexity, a scan of the section headings and a look at the conclusions. This would probably be followed by a reading of carefully selected sections or passages.

A major problem facing the speaker therefore is the audience response in a situation where he has taken away its freedom of action. He has to make the decisions with regard to order of presentation of information, the speed of presentation and the need for any recapitulation. His words are not permanent and available for reference. Once spoken they are gone, and if the message has not been communicated within a few seconds then it is lost forever as far as the audience is concerned.

4   *Preparation of Material*

In order to provide a general account of how to prepare material for presentation it is supposed in this chapter that there is ample time available. The question of preparation when time is very limited or approaches zero is discussed in Chapter 6.

4.1   THE FULL TEXT

The first stage in preparing your paper or talk is similar to that which you adopt in writing reports. Collect together the ideas and information that you want to communicate and arrange them logically. Then write out in full what you think you would like to say. If time permits it is always best to start with a full text. Thoughts are not complete, and may not even be true or logical, until they have been expressed in the careful language that a written text demands.

You will find at this stage that what you have written is very similar to a technical report. It will not be suitable for verbal presentation in this form, bearing in mind what we have said earlier, but it is the best starting point. The next stage is to revise the text with regard to order, style and length.

Order

The order in which you will have assembled your ideas will probably follow a conventional report format. This is excellent for a document that needs to be referred to: the reader knows where to look for the information he wants. It is however not very relevant for a verbal presentation.

The opening is important. Avoid leading into the subject too conventionally or too leisurely. The purpose of the opening is to catch the attention. If you can catch the attention immediately you have at least a chance of holding it. If you bore everyone in the first minute with information that could be taken for granted, you make it harder for yourself when you come to the important parts. Examples of the kind of openings to be avoided are as follows.

'Thank you Mr Chairman for those very kind words. It is a great pleasure for me to be here once again in this great centre of learning. . . .'

'This paper is based on investigations carried out in the Department of Pure Mathematics at the University of Cambridge. . . .'

'Dr Watson, my co-author, and I discussed who should present this paper and we decided that I should take on the task. Dr Watson is here with me on the platform and will assist in answering any questions that . . .'

The information contained in these examples may be worth giving, but not in the opening sentences. The opening is too valuable to be wasted in this way. Try to extract a theme, a key message, from your material and word it in a way that gives impact. Consider the following examples.

'Was it possible to increase the efficiency of a Z500 water pump. That question . . .'

'The fracture characteristics of various oxides of iron can be explained in terms of . . .'

'Practical use of stress analysis can advance no more quickly than the solution of stress problems by advanced mathematical techniques.'

It is permissible to use a non-technical opening provided it has a link with the subject under discussion. Consider the opening of a presentation of a paper on corrosion with the following words taken from the Bible. 'Lay not up for yourselves treasures on Earth for the rust and the moth to consume.' The main criterion in selecting an opening should be 'Does it catch the attention?'

Following the opening comes the body of the paper. Your task here is to hold attention remembering the shape of the attention curve discussed earlier. Experiment with changing the order of the ideas. Don't be too concerned with the chronological order of events. Think of the effective way attention is maintained in novels, films and plays by the use of flashback and change of scene. A way of raising the attention curve is to summarise what has been said so far. Repetition is not permitted in written reports but it is permissible and desirable to a limited extent in verbal presentations. It is possible to maintain attention during fairly lengthy speeches by periodic revision, each revision linking the new material with that preceding the last revision.

Obviously, it is important not to say anything that will cause an immediate loss of attention. If you say 'This is not important but . . .' or 'I don't wish to deal with this but . . .', you will invite your audience to take a mental rest. You then have the difficult task of indicating that the rest period is at an end and that you are again talking about what is important and what you do intend to deal with. A commonly heard phrase is

'I'll cover that later . . .' or 'I'll go into that in some detail later
. . . .' This comes over as a promise of heavy going, and if
such a promise is made several times the audience becomes
apprehensive about how much longer the speaker intends to
be and how much more detail he is going to try to squeeze
in.

Loss of interest can also be caused by poor linking phrases
such as 'Next, I want to talk about . . .' or 'which reminds me
of . . . .' Links should be carefully planned to give a smooth
transfer from one idea to the next. That is not to say you
cannot use contrast, but you must lead into it effectively. 'An
entirely different theory is . . .' or 'On the other hand . . .' are
examples of satisfactory link phrases.

The final part of the presentation is the conclusion. This is
the word commonly used. Unfortunately there is a danger of
identifying this with the conclusion of a written technical
report. The latter is conventionally a statement of the main
findings of the work described. The conclusion of a speech is
the end part. It may of course be appropriate to use the
technical conclusions as the conclusion of the presentation,
but be prepared to accept that this need not be the case.

The conclusion is important because it is possible to achieve
increased audience attention towards the end of your speech
(see the attention curve in Chapter 3). It is also possible to
achieve very high audience retention of the final statements.
Clearly you need to signal that the conclusion is coming.
If you say 'And finally, Ladies and Gentlemen . . .' or 'In
conclusion . . .' there will be an awakening of interest. Those
who have lost track of your argument or simply tired in the
heat of the afternoon will make an effort for two reasons.
Firstly they think that if they grasp your conclusion they will
have at least something to take away, and secondly they expect
their effort to be of limited duration because you have
indicated that you are about to finish. However, be careful to
avoid false conclusions. If you have already said 'and
finally . . .' three times don't be surprised if some of the
audience are already chatting among themselves.

In selecting your conclusion decide on the message that you would most like your audience to remember. It may be a specific technical conclusion or it may be much more general. Consider the progression in the following list of examples.

'Galvanising of structural steelwork to a thickness of 100 μm will give protection in a tropical environment for about two years.'

'Techniques are now available for measuring the protective ability of galvanising in tropical environments.'

'The use of structural steelwork in tropical environments gives rise to expensive problems and we must continue to develop and improve methods of evaluating the protection given by galvanising and other treatments.'

'One of the key factors in industrial development in tropical areas is the greater use of structural steelwork. We have to ensure adequate life of the steelwork by effective use of protective systems.'

It is a good idea to consider the conclusion and the opening together. If your conclusion links back to the opening, your audience will have a pleasant feeling of tidiness and completeness. Your message will have been neatly packaged for taking home.

Much of what I have said about the structure of your presentation is embodied in frequently quoted principles of speech preparation. The opening is to catch (the attention), the body is to convince and the conclusion is to confirm. The idea is alternatively expressed as 'tell 'em what you are going to tell 'em, tell 'em, and tell 'em what you've told 'em.'

Style

The second way in which your full text needs modifying is with respect to style. This is where you have to put aside the years of training and practice in preparing written reports.

Write out the text as you would speak to someone in your

natural spoken style. Avoid long complicated sentences. We rarely use them in conversation. Punctuation is a device introduced to make written passages intelligible. It is not a feature of verbal communication. In speaking we use pauses and variations in pace and tone of voice to make a sequence of words carry meaning. These characteristics are not evident when we write the words so we have to use punctuation to make the meaning clear. When you write your text therefore you should not feel obliged to punctuate in any conventional way. One of the most useful symbols to use, for example, is a dash (—) which can indicate a pause. The pace and tone variations are known to you as you compose and you can use personal coded symbols if you wish to note them.

To some extent grammar can be considered in a similar way to punctuation. Verbal communication existed long before grammar. Grammar is a set of rules extracted from usage initially but then applied to subsequent usage in a praise-worthy attempt to standardise. The need for rules of grammar, however, arose mainly because of the use of written communication. If the spoken word had remained our only means of communication it is doubtful whether we would have ever seen the need to rationalise grammar. Speech develops empirically according to its effectiveness. Therefore, write out your text as you would say it to give the greatest impact and interest. Remember that much of what we say is not in the form of sentences. We use exclamations and phrases that would be inadequate and irritating if written, but with the right voice they can be full of meaning and interest.

When revising the style of your speech remember that the atmosphere in which you are making the changes is different from the atmosphere in which you will make your presentation. Alone in your office or study you will feel supremely confident and enthusiastic about your material. This can give rise to a style of composition which before an audience will sound superior or even pompous. As you write, constantly visualise the faces of the audience. Remind yourself that your listeners, perhaps not experts in your subject, are nevertheless capable of critical thought and evaluation of your messages.

Length

The final process of modification of your full text is adjustment of the length, and this will almost always mean shortening.

How much time is available for your presentation? This may have been imposed on you or you may have some choice, but either way you must be certain how many minutes you intend to speak for. Count the number of words in your text and assume a presentation rate of 100 words per minute. Some would consider this to be a maximum and would prefer 80 words per minute particularly for a speech that is to be presented in a large hall. I consider however that 100 words per minute is probably the best value to work to for the range of circumstances likely to be encountered by the average scientist or engineer.

Your immediate reaction will be that 100 words per minute is too slow. The value is, however, based on the rate at which members of an audience can absorb new information. If you do insist on going faster, then only part of what you say will be absorbed. I know from my own experience that people are not easily convinced of this proposition, until they have put it into practice and observed the effect. Opposition to the concept of not more than 100 words per minute arises because most speech does proceed at a greater rate. In conversation, for example, we often speak at twice this rate. In conversation, however, most of us repeat everything at least once. We only listen to part of what is said and we ask questions until the message is understood. There are of course circumstances in which we can absorb information at rapid rates. If we are familiar with the material to some extent, if we concentrate intently for a short period of time, if our need to obtain the information is very great or if the information is trivial or simple in content, then we can beat the 100 words per minute barrier. Such circumstances do not generally apply to audiences.

You must shorten your text if it is too long. You must not overrun your allotted time. I know that most speakers do overrun their time but you must accept as your primary rule that you will not. Great importance should be attached to this for the following reason. Even if your presentation turns out to be not so well executed, the audience's estimation of your performance will be greatly enhanced if you at least keep to time. If you overrun with a poor presentation you are adding insult to injury. Speeches that extend beyond their alloted time irritate the audience, other speakers, the chairman and the organisers. Once the hands of the clock have passed the planned completion time, few are listening with any interest and the impact of the speaker's conclusion is lost. The audience is filled with thoughts of the coffee break or missed trains. The chairman is wondering whether to stop the speaker or whether to pretend everything is still under control. Other speakers waiting their turn are feeling robbed or planning themselves to overrun in order to extract retribution.

Let us consider how you should undertake the shortening of your text. Recognise first that there is no point in shortening it by economising on words. The use of less simple words or clever syntax will no doubt reduce the word count but this is delusory. The word count is simply an arbitrary and approximate measure of the thought content that has to be absorbed by the audience. Furthermore the words in your text will not be the actual words that you use in your presentation as we shall see in the next chapter.

Work backwards in the sense of asking yourself what the basic message is that you wish to put over. Then decide on the minimum information that you need to use to justify the basic message. Anything left is not essential.

If you still have a problem at this stage with regard to length you should consider the following possibilities. If your presentation is of a paper at a conference, the audience may have been supplied with a preprint. You can refer to the fact that certain details which you do not propose to discuss are given in the paper. You can even make similar reference if the paper has yet to be published. If your presentation is to be at a meeting of a committee you can table a document which provides details that you can then omit in your verbal presentation. In many circumstances you can use a visual aid (dealt with in Chapter 7) to economise on explanation. Another technique you can use is to simplify the arguments, but making it clear that you are doing so.

4.2   NOTES

It may come as a disappointment to learn that the full text, by now so carefully prepared, is not going to be used for the actual presentation. From the full text suitable notes need to be constructed. There may be circumstances which demand the reading of the full text, and these will be mentioned later, but accept as a general principle that you should not read a speech to an audience but should work from notes.

The first thing to do is to read through the full text underlining key words or phrases in each passage. Use a pencil so that changes can readily be made. List the underlined words and phrases in order on a separate sheet of paper and then attempt to speak aloud the content of the full text using the list as the only prompt. Notice that the idea is not to attempt to remember the exact words of the full text. If the prompt brings to mind the idea that has to be expressed then it is serving its purpose. Do not concern yourself with the particular words you are using at this stage.

The initial listing will be found to be unsatisfactory in two ways. First, some of the words and phrases will be unnecessary, the previous prompt being adequate to give recall for more than was initially expected. The prompts do not have to be spaced evenly through the full text. Sometimes a single word will trigger off a long passage and sometimes a lengthy phrase may be needed for a single sentence. The opening and the conclusion will probably need more detailed notes than the body of the speech.

The second kind of difficulty is that the selected phrase does not trigger the correct idea. This means that a change is required. Look back at the full text. Almost certainly you will, in selecting your key words and phrases, have used mainly nouns. These are not necessarily the best words to use. The names of things, pieces of apparatus or places for example, will have to appear in your notes but most nouns relevant to your subject will be recalled very easily. You have, after all, been working with these things day after day. Sometimes the best key words are words like 'nevertheless', 'initially' or 'however'. Such words trigger the correct mood for what is to be said. Nouns will trigger only the content.

Experiment with changes in the key words and phrases until the list is satisfactory in triggering the content of the full text and the atmosphere in which each idea is to be presented. Make use of symbols where necessary. I have found small round faces drawn in the margin very helpful. These can be shown smiling to warn of a light hearted comment that is due, or with a solemn expression indicating the need for a slow, serious, well emphasised statement. Notes are for personal use only and you should not be afraid to develop a personal system which suits your needs.

When you are satisfied with the list of key words, phrases and any other marks or symbols, return to the full text and underline the corresponding words and phrases in red, preferably with a felt tipped pen. Add your marks and symbols in the margin again in red.

Now prepare your list, or notes as it has become, in the form in which you will use it for the presentation. Write out your notes on cards, not paper. Postcards or file cards of a similar size are ideal. File cards can be obtained in a variety of colours. As an alternative to white which is conspicuous consider using blue, say, to blend in better with the clothes you will be wearing. Write across the smaller dimension so that the longer sides of the cards are vertical. Write on one side only and number each card in sequence at the bottom. Use only the upper three quarters of each card. The lower quarter should be empty, apart from the card number, but you can usefully include a reference or paper title at the bottom in case the notes are required at some later date. It does not matter how many cards you use, simply continue adding as required.

Paper should not be used for notes. It bends, moves about, makes rustling noises and sticks to the next sheet. One side only of each card should be written on, otherwise you will never know when to turn over and when to move the card to the bottom of the pack. You are likely to remind the audience of a conjuror who is trying to do the three-card trick. The reason for leaving the bottom of the card blank is so that your thumb can locate itself there. I have spoken in terms of 'writing' notes and it is in fact better not to have them typed. Typescript is rather small for notes. It is better to write with letters large enough for you readily to see what is written at a glance. Notes prepared in this way can be conveniently carried in pocket or handbag. This allows a glance through at odd spare moments and while waiting to present your speech.

As a final operation, punch a hole through the bottom corner of each card and link the pack together with a treasury tag. This will keep them in the correct order.

4.3   PRACTICE

Read through the full text to bring to mind what you intend to say. Now take up a position in front of a mirror armed with your notes. Stand far enough back so that you can see your head and feet. It is better to stand than to sit because if you can get used to being relaxed when speaking while standing you will have no problem when you have to speak from a sitting position. Place the feet slightly apart with one slightly in front of the other. This will help to inhibit any tendency to sway from side to side or front to back.

Hold your notes in one hand and let the other hand hang limp at your side. The notes should be held high and forward so that you do not have to bend your head downwards through a great angle each time you wish to refer to them. Obviously the notes should not be so high that they cover the face or so far forward that you give the impression that you have forgotten to bring your reading glasses.

Examine your reflection in the mirror for a few seconds. You should look stable, relaxed and natural. Notice and remember that you look better than you feel. That arm in particular, the one hanging by your side, feels strange but in fact looks perfectly natural. Getting used to a suitable stance is one of the problems that most speakers have difficulty with. Some practice is needed and more is said about this in Chapter 9.

Now take two slow, fairly deep breaths, say 'Mr Chairman, Ladies and Gentlemen', pause, then start. Look at your notes as you need to but when you have mentally grasped the prompt look at your reflection and speak. Avoid looking at your notes when you do not need to. Observe your reflection, watching for any swaying, rocking or waving of notes. Make sure your spare hand does not wander into your trouser pocket and start counting your loose change.

The words you will be using as you proceed will not be the same as the words in your full written text. This is as it should be. Do not attempt to memorise the full text. Some phrases will unconsciously be remembered and used, but they will have the conviction of the moment. This is what gives a speech life and holds the attention of an audience more than anything else. The words you speak should be full of feeling and conviction of their meaning. The feeling and conviction should not be remembered emotions but ones that are present in your mind as you speak the words. Your concentration therefore is on ideas rather than words. The notes prompt the idea, the idea must be nourished by feeling so that it gives birth to words. If the idea and the feeling are correct the words will be exactly right. Worry about the idea and the words will look after themselves.

The intense awareness necessary to speak effectively will

arise more readily if you let the notes form pictures in your mind. Visualise, then describe what you see.

Memorised speeches lack conviction, but there are other reasons why speeches should not be memorised. If you rely on memory there is the danger of a distraction or disturbance causing you to momentarily forget where you are. It can then be very difficult to pick up the thread again without recapping or missing something out. A further problem with a memorised speech is that last-minute changes are not possible. You may find as a result of something a previous speaker has said you need to amend a statement in your speech. Or you may wish to include a comment relating to a point that has arisen a few minutes previously. In such circumstances it is impossible to make appropriate changes or insertions if you are relying on a memorised version of your speech.

You will probably wish to try one or two more practice runs. You should time the presentation and slow down as necessary. I say 'slow down' because it is fairly certain that you will have spoken too quickly initially. A tape recorder is very useful in allowing you to analyse your performance in some detail. Notice any 'um's or 'er's. Removing these takes practice but being aware of them is a first step in the right direction. Listen particularly for lack of speed variations or tone variations. To avoid a monotonous speech you have to use such variations. They arise essentially from conviction. As discussed above, each statement must be spoken with a totally dedicated mental feeling for the meaning of the statement.

All that now remains to be done is to make any final changes that seem appropriate. However, avoid making too many changes at too late a stage. There is a danger that during your actual presentation you will divert from version number four into version number three.

4.4   EXAMPLE OF SPEECH PREPARATION

In this section an example will be given of the preparation of a verbal presentation. First an initial draft of a full text will be shown. In order to avoid copyright problems or letters from enraged authors I am using a hypothetical paper which describes a fairly simple school experiment in physics. The text is short but sufficient to illustrate the various points that have been made. Then follows the rewrite of the draft. This is the essence of what the speaker will say. Finally the notes used for the verbal presentation are reproduced.

Full text (initial draft)
Introduction. The purpose of the investigation to be described was to examine the relationship between the pressure and temperature of a constant volume of gas. Previous workers have shown that the pressure increases with temperature. This result is not unexpected. Since expansion is expected as a result of temperature increase it would follow that prevention of such expansion would give rise to an increase in pressure.

The present work reveals for the first time the quantitative relationship linking pressure and temperature. The paper discusses the possible implications of such a relationship and draws attention to practical applications.

Apparatus. The apparatus consisted of a glass bulb joined to a capillary tube which was connected to a mercury reservoir by a length of rubber tubing (Fig. 1). By means of a three-way tap, air could be pumped from or admitted to the glass bulb, which was mounted in a water bath. The mercury reservoir was mounted on a vertically moving slide which could be accurately positioned. A scale allowed the mercury levels in the reservoir and connecting tube to be determined.

Experimental Procedure. Dry air was admitted to the glass bulb. The temperature of the water in the bath was observed after stirring. The level of the mercury in the connecting tube and reservoir was observed on the scale.

The temperature of the water was then gradually increased. This caused the mercury level in the reservoir to rise and that in the connecting tube to fall. The reservoir was moved upwards on the slide so as to bring the level in the connecting tube back to its original position. At 10 °C intervals the temperature of the water and the level of the mercury in the reservoir were noted, care being taken to ensure that the mercury level in the connecting tube was at its original position. The procedure was continued until the temperature of the water reached 100 °C. The value of atmospheric air pressure was determined by means of a Fortin barometer.

Results. Because the level of mercury in the connecting tube was maintained at a constant value, the volume of air in the glass bulb remained constant. Thus the pressure as indicated by the difference in the two mercury levels was the pressure resulting from the temperature increase of the fixed volume of air. The measured values of pressure were expressed as absolute values by adding the value of atmospheric pressure.

The results are shown in Fig. 2 where the values of pressure are plotted against the corresponding values of temperature. It can be seen that there is a linear relationship. The slope of the line was found to be such that it would, when extrapolated, intercept the temperature axis at −273 °C.

Discussion. The observation of a linear relationship is of particular interest since, if linear extrapolation is justified, the result would predict zero pressure at a temperature of −273 °C. Such considerations are purely speculative of course. Air is known to liquefy well before reaching such a low temperature. Even if this were not so it is probable that the linear relationship would give way to a curve and approach the temperature axis asymptotically. In this way the difficulty of envisaging zero pressure at a particular temperature would be avoided.

Further work is clearly called for. Similar experiments with other gases are planned and consideration is being given to an improved apparatus which will allow observations to be made over a wider temperature range.

The practical importance of the investigation must be emphasised. The data allow design calculations to be made for machines using heated air as a power source. Furthermore the apparatus forms the basis for the design of an accurate thermometer.

Conclusion. A novel apparatus was used to study the temperature–pressure relationship for air at constant volume and showed that the relationship was linear over the range 20 °C to 100 °C. The results were found to give an extrapolated value of zero pressure at a temperature of –273 °C, but the full implication of this observation is not yet clear.

(The text contains 670 words. There are 37 sentences giving an average of about 18 words per sentence.)

Full text (final version)
If you heat a quantity of gas it will expand. If you resist the expansion and maintain the volume of the gas constant then the pressure increases. A question of great practical importance—'What pressure can be generated by a given increase in temperature?' This is the question that we set about answering.

The basis of our method was to contain a quantity of dry air in a glass bulb. We heated the air by immersing the bulb in a water bath, and we made our observations at 10 degree intervals from 20 °C to 100 °C. The first slide shows the arrangement we used to maintain the volume of air constant. A mercury reservoir is connected to the bulb. As we raised the reservoir we opposed the increase in air pressure in the bulb. In this way we could restore the mercury level in the connecting tube to its original value. The difference in height between the two mercury levels gave us the value of the pressure increase. To this we added the value of atmospheric pressure measured at the same time.

The next slide shows the results we got. We plotted pressure against temperature and got a straight line. This data relates to air and will be valuable to machine designers. We plan to repeat the experiment with other gases. One thing we particularly want to do is to extend the line over a wider temperature range. For this we need an improved apparatus.

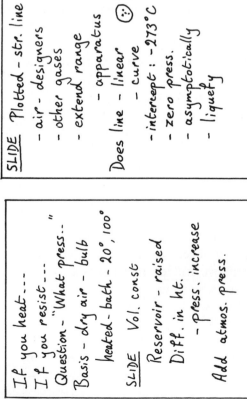

Figure 2   Example of speech notes prepared on 5 in × 3 in file cards, linked together by a treasury tag inserted through the hole in the bottom left-hand corner.

Does the line remain linear? Does it start to curve as it approaches the temperature axis? If it were to continue at the same slope it would in fact intercept the temperature axis at –273 °C. Thus the pressure would be zero at this temperature. It is much more likely that the line would approach the axis asymptotically so that zero pressure could never be reached. We know in any case that air will liquefy before it gets anywhere near −273 °C.

In conclusion, therefore, can I summarise by saying that we have provided basic pressure–temperature data for air over the range 20 °C to 100 °C. This will be of value to machine designers. The linear relationship we have observed prompts some interesting questions and indicates the need for further investigations. Some of you may have noticed a possible spin-off from our investigation. Our apparatus could form the basis of an accurate thermometer—we are working on it!

(The speech contains 400 words and is therefore suitable for a presentation time of 4 minutes. There are 31 statements giving an average of about 13 words per statement).

Notes
The notes derived from the underlined parts of the above text are shown in figure 2. They are shown handwritten as they would be on cards of postcard size.

5   Presentation

5.1   PRELIMINARIES

Before the actual moment of starting to present your speech there are a number of preliminary tasks to be carried out. In a sense they form part of the preparation stage but they have been included in this chapter because they are usually carried out not long before the presentation.

Spend a few minutes in the room in which you are to make your speech. Walk round and view it from different angles to get a general feeling of the atmosphere. Find out where you will be expected to stand while speaking. I am assuming for the moment that you will be standing. This is the usual arrangement at a conference for example, but there are some circumstances such as committee meetings where it would be normal to speak while seated. The preliminaries relating to committee meetings and similar occasions are dealt with in Part III. It is always better to speak standing if you have any choice and if circumstances allow. Your voice will project better and you will hold the interest of the audience more easily.

If you have decided to use any visual aids in your presentation (see Chapter 7) or if you will be using a microphone, then the position from which you speak will probably be determined. Avoid the use of a lectern if possible. A lectern presents a psychological barrier between the speaker and the audience and makes the speaker's task of persuasion more difficult. A lectern can become a leaning post for the speaker who gradually progresses from resting his hands, then elbows on the welcome support, to posing like the Hunchback of Notre Dame. It is much more effective to stand so that the audience can get a full-length view of you.

Take up your position and try your voice. 'Mr Chairman, Ladies and Gentlemen.' Listen to the resonance but remember that with the audience in position the sound will be deadened. Imagine there is someone sitting on the back row and speak so that he will hear you.

Try the microphone if you are to use one. Find out where the switch is and which way is on and which way is off. Move your head about as you speak and notice how directional or otherwise the microphone is. Notice the effect of varying the distance between your mouth and the microphone.

If you are using visual aids make sure you know how to operate everything. Try switching the overhead projector on and off and changing the focus. In addition, on a self-operated slide projector, you need to know how to advance or retrace from slide to slide. Make sure you know where the chalk or flip-chart pens are.

A particularly important job is to plan your movements. You may, for example, have to walk from the microphone to the overhead projector or blackboard. It may therefore not be possible to speak while you are presenting a picture or diagram. You must return to the microphone before continuing otherwise you will irritate the audience at best and have parts of your speech unheard at worst. Provided you have recognised problems such as this in advance they will not trouble you.

If time permits, a full rehearsal in the room is ideal, but if this is not possible you must ensure that you have at least a few minutes in the room to check on the essentials.

As you have been reading through this chapter you have no doubt been thinking that these preliminary checks are fairly obvious. You have probably decided that this is the procedure you will adopt and that the matter needs no further consideration. Let me give a word of warning. The reason why many speakers do not carry out these preliminaries is not because they do not accept the value of them. Nor is it necessarily because they have not resolved to carry them out. The principal reason is that when the time comes to carry them out, the speaker feels he should not appear to be inadequate. He feels that he must appear to be casual, to have everything under control and to be supremely confident. The last thing he wants to do is to appear to be 'fussing', worrying unnecessarily or pestering the administrators with questions about the microphone or projector. After all, none of the other speakers are worrying about where the microphone switch is.

It is not sufficient therefore to merely acknowledge the value of the preliminary checks and to decide to carry them out. It is necessary for you to resolve in advance to carry them out even if, when the time comes, you would prefer not to.

5.2   INTO BATTLE

As you sit awaiting your turn to speak you can feel your heart beating faster than normal. Your stomach quivers with 'butterflies'. Your mouth may feel dry, or, alternatively may be producing an excess of saliva. The room feels warm: you may be breaking out in a sweat. You have a repeated desire to swallow. All these symptoms are perfectly normal. Your body is preparing itself to deal with an important event demanding your full concentration and capabilities. This is what is happening and this is what you must tell yourself is happening. Accept the feelings as a natural process; let them happen. You know that these feelings will come under control when you start speaking. You know that is so, because you have done your preparation properly.

Listen carefully as you are being introduced in case the chairman makes a serious mistake in saying who you are or what you are going to talk about. Note the time and check that your clothing has not become disarranged. Buttons, zip fasteners, ties and pocket flaps represent the main hazards.

When the introduction ends take up your speaking position. Stand holding your notes in the way that you practised. Resist the temptation to start too quickly. Take two slow fairly deep breaths. In . . . out . . . in and 'Mr Chairman, Ladies and Gentlemen.' Then pause. These opening words, or their equivalent in other circumstances, are important. They allow you to test your voice, to test the microphone if you are using one and to test the acoustics of the room with the audience present. Listen to the resonance. Look around while you pause and judge whether everyone has heard you clearly and is paying attention. If you consider that the opening statement was not satisfactory you can correct as you proceed. The point about the formal opening is that it is not part of the speech. The information contained in it is irrelevant. No harm has been done if you have not been properly heard.

You will notice that I included testing the microphone in the previous paragraph when discussing preliminary tasks to be carried out before the meeting. That is the proper time to test the microphone. Never tap or blow down the microphone to test it when you are about to start your presentation. This is only one stage removed from the 'testing, one, two, three . . .' syndrome. Not only is it irritating to the audience but it gives the impression that you do not know whether the microphone is working or not. If you have carried out your preliminary checks then you will know that the microphone is switched on. Give the audience the impression that you do know it is switched on. Usually you will be correct. If something has gone wrong then you have not ruined the opening of your speech. You have merely lost the formal address to the chairman and the audience, which can readily be repeated when things have been put right.

Having paused you are now ready to start your speech, correcting as necessary the speed or volume on the basis of the feedback that you have just assimilated.

At this moment there will be a strong temptation to depart from your notes. Resist it. Do not thank the Chairman or thank the audience. Do not attempt to tell them that you are pleased to be there, and above all do not elaborate on your lack of skill as a public speaker. Audiences do not like speakers to apologise. They find it embarrassing. They are basically on the speaker's side: they want him to get on with what he is there to do and they want him to succeed.

Glance at your notes and absorb the meaning of your opening statement. Look at the audience and speak with intense awareness of the meaning that you wish to convey to them in your words. Concentrate on the meaning, the words will look after themselves. Glance at your notes again and continue in the same fashion. As you speak look at the audience and stop speaking when you look at your notes. Direct your gaze to the back of the audience rather than the front row. Look across the audience from side to side. Avoid picking on some poor fellow and looking at him to the exclusion of others. You will make him feel uncomfortable and he will begin to wonder if he has a smudge on the end of his nose. If you find it slightly disturbing to catch the eyes of individuals, avoid this by focusing just above the heads of the audience.

In looking round the audience be careful that the sound does not fade as your mouth rotates away from the microphone. Move your head so that the mouth is the pivot point. A particularly common and bad variant of this is to produce fading of sound by turning to look at the screen or blackboard. You should look at your visual displays only when it is necessary and you should not speak while looking at them.

As you speak you may find that your style is developing an over-modest characteristic. Watch out for it and recognise its origin. It is evidenced by excessive uses of phrases such as '. . . you will have seen this kind of thing before . . .', '. . . there is nothing special about this, but . . .', '. . . some of you will know more about this than I do . . .', and so on. Such phrases are permissible when they are in fact true. It is their excessive use that you must notice. The fault is particularly hazardous because it will not have featured in your practice sessions. In the security of your office or home you will have felt sincerely that the information you are to present is worth presenting. During the actual presentation, however, nervousness can show up as a lack of faith in your material. You will feel that your information is inadequate, that your results are not as good as they should have been, that the members of the audience have heard it all before or even that they are secretly laughing at your presentation. Such feelings are illusory, (provided of course that your preparation has been correctly carried out) and you should recognise them as another symptom of nervousness.

The way to cure this problem is to fix firmly in your mind the fact that you have carried out your preparation properly. Your material is therefore correct for the occasion. 'In any case,' you must tell yourself, 'rightly or wrongly my material is determined once I start my presentation. My purpose during the presentation is to do justice to my material as planned, not to have second thoughts about it.' Continue speaking with the intense awareness of the meaning of your message, as has been described previously, and couple this awareness with a renewed enthusiasm for the value of your message.

When you come to the end of your speech resist the temptation to add something extra. Do not thank the audience and do not offer any apologies of any kind. Your prepared ending is your final statement. The ending is signalled by the way you make the final statement, by the tone of voice and the pace. To get it right you have to be intensely aware that this is the ending. You must say it as if you will never again speak to these people and you want above all else that they will remember these final words to their dying days.

Pause after delivering your final word and then move from your position back to your seat to the sound of deafening applause!

6   Speaking with Limited or no Preparation

There is no substitute for thorough preparation. It is even possible to compensate for lack of experience by spending extra time in preparation. Nevertheless there are circumstances in which you will have to speak without having had the opportunity to follow the procedure detailed in Chapter 4.

This chapter will give you techniques to apply in such circumstances, but always bear in mind that lack of experience reveals itself more readily when preparation time is short. If you wish to develop a skill for speaking with limited preparation, or with none at all, then you should consider devoting a little time to the exercises designed for this purpose which are described in Chapter 14.

6.1   READING A FULL TEXT

As mentioned previously you should avoid reading a full text. Some of the worst presentations are the result of speakers reading papers from beginning to end. Life is busy, however, and time is at a premium. There are situations in which you will have no choice. A document will arrive at the last minute and you will be forced to read it.

There are also situations in which you cannot afford to deviate from the prepared text. A carefully worded statement of policy, terms of reference of a contract, an observer's description of an event under investigation, a statement involving exact values or sums of money: these are such examples. Generally this will not apply to an entire document but rather to parts of it.

The main rule to follow in reading, whether it be a complete paper or extracts, is to read with visual communication. This means that you must look at the audience as you speak the words. You are, of course, permitted to glance at the text, but to stare at it for long periods is to isolate yourself from the audience and lose their interest. I know it sounds impossible to do but it is possible. However, it is not easy to do without practice.

Before starting, read through the text and using a pen mark off each short phrase with a stroke. You can use a double stroke where the sense is complete. This will be so at each full stop but perhaps elsewhere in addition. To illustrate the method the beginning of the example at the end of Chapter 4 can be marked as follows:

The purpose / of the investigation / to be described / was to examine / the relationship / between the pressure and tempera-ture / of a constant volume / of gas. // Previous workers / have shown / that the pressure / increases with temperature. //

Hold the text in your left hand. If it is on sheets of paper use a board or a piece of stiff card at the back to give support. Now place the index finger of your right hand under the first phrase 'The purpose'. You are going to use this finger to keep your place. You were taught the technique when you first learned how to read at infant school but, unfortunately perhaps, you dropped the habit because of its childish connotations.

Glance at the words 'The purpose', look at the audience and speak the words. As you do so, move your finger to the next group of words. Glance down, look at the audience and speak 'of the investigation.' Continue in this fashion at a fairly slow pace. Remember the 100 words per minute rule. If you have held your text high as we discussed in Chapter 4 you will have discovered that it is hardly necessary to move the head as your eyes move between the text and the audience.

With a little practice you will find it possible to give a smooth and intelligible reading of a text, even one which you are unfamiliar with. You will find that in your downward glances you are looking ahead to get the sense of what is to come. The main fault to be aware of is using a descending tone of voice at the end of each phrase. This will close the sense and produce a staccato effect. In your initial marking of the text try to spot places where there is a false closing of the sense signalled. For example, in

'The purpose / of the investigation / to be described / was to examine / , , ,'

notice how a closing of the sense after 'investigation' will leave 'to be described' suspended in mid-air. To ensure that you keep the sense open put a suitable mark over the word investigation. Thus,

'The purpose / of the investigation / to be described / was to examine / . . .'

If you have time to have the paper retyped before your presentation you can use a layout as shown below. This makes it easier for you to follow. It is rather wasteful in terms of paper but paper is cheap compared with your reputation!

The purpose
  of the investigation
    to be described
      was to examine
        the relationship
          between the pressure and temperature
            of a constant volume
              of gas.

Previous workers
  have shown
    that the pressure
      increases with temperature.

6.2 SPEAKING FROM A FULL TEXT

It is better to speak from a full text, using words of the
moment and a spoken style than it is to read a full text. In
order to do this, it is necessary to be to some extent familiar
with the content of the document and to have at least a few
minutes to make your preparations.

Prepare the full text by underlining key phrases or words in
red. This you will recall was the first stage in preparing notes
from a full text. The underlined words can now be used as
notes for the presentation. As an alternative to underlining
you can use a pale-coloured broad-tipped felt pen to paint
across the key phrases.

The main pitfall to avoid is glancing at other parts of the
text that are not underlined. Such glances will hinder the
composition of your own statements. Get used to the technique of glancing at the prompt, absorbing the sense and
speaking with conviction of the meaning. If the prompt is
inadequate and you find yourself in difficulty then read the
statement with visual communication as described in the
previous section.

With skill it is possible to transfer smoothly from speaking
from the prompts to reading with visual communication, and
back again. The technique can be quite effective. You must
always be clear in your mind, however, at any moment as to
whether you are doing one or the other. Any confusion will
cause you to fail to do either.

6.3 SPEAKING AT SHORT NOTICE

There are many occasions when a speaker has insufficient
notice to even contemplate the possibility of writing out a full

text. This is what I mean by speaking at short notice. Essentially the task is to prepare notes direct from ideas.

The instinctive approach is to start making a list. There is a better method. Tony Buzan[4] devised what he calls 'Brain patterns' and, although he does not appear to be aware of it, these provide an excellent means of collecting and organising ideas for speeches. It is worth referring to his book to see the theory behind brain patterns and to appreciate their use in other applications such as study, essay writing and note taking. For present purposes it will suffice to describe how to use a brain pattern to prepare notes for a speech.

Take a sheet of paper and write the theme of your speech in the centre. Encircle it. Draw lines branching from the circle each representing an associated idea labelled appropriately. Any branch can then give rise to further branches as required. The advantage of such a network is that any idea as it arises can be located in a logical position. The multiple branching of the network keeps all the concepts under review at the same time and shows interconnections. A list on the other hand is inhibiting. It concentrates attention on the last item and sends the thinking processes down one path. It fails to show inter-relationships between the various items.

Several examples of brain patterns are shown in figures 3 to 6.

Having produced your brain pattern decide on your basic message. This means concentrating on some branches and merely mentioning or leaving out others altogether. Underline or ring the main items and note any interconnections. Now decide on your conclusion. Remember what has been said previously about conclusions. The conclusion should be the message above all others that you wish your audience to take away and remember. Decide on an opening which will catch the attention and plan your route through the brain pattern. Finally, list the opening statement, the items in order, and your ending. This list is now your notes. Do not worry about, or waste time composing, the words you will actually use. Concentrate on ideas and the words will look after themselves.

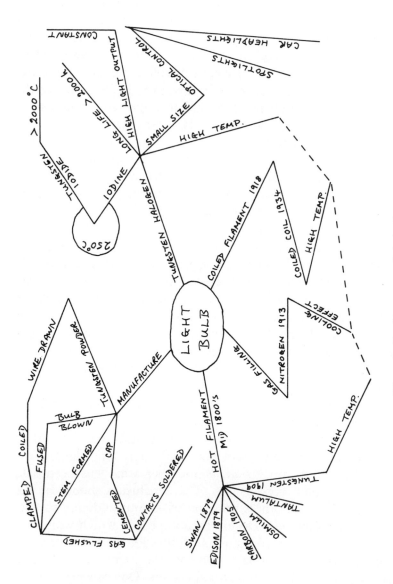

Figure 3   Brain pattern—light bulb.

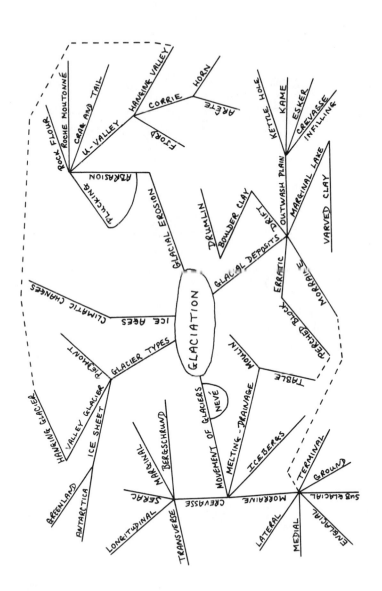

Figure 4 Brain pattern—glaciation.

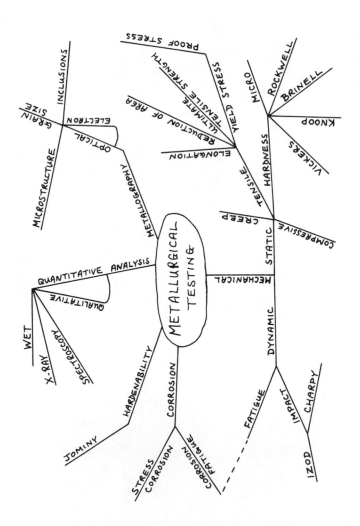

Figure 5   Brain pattern—metallurgical testing.

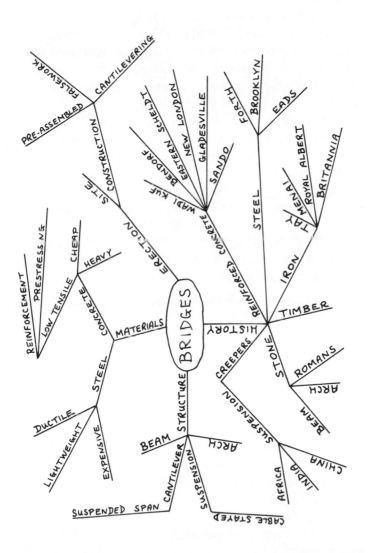

Figure 6 Brain pattern—bridges.

You may not have time to have a practice run through so timing may be a problem. Almost certainly you will have too much material for the permitted time and equally certainly you will think you have too little. To help you begin to get a feel for time think of the following. A five minute presentation requires no more than 500 words. A statement will need about 12 words, so you have time for no more than about 40 statements. With three or four statements per idea you will cover no more than about a dozen ideas.

By way of example, let us suppose that you have to speak at short notice for five minutes on light bulbs. We shall suppose of course that you are knowledgeable on the subject. A brain pattern that you might well construct is shown in figure 3.

On reviewing the brain pattern you decide that as time is restricted you cannot cover the manufacturing processes. You will adopt as your basic message the important new development of the tungsten halogen bulb and will set it in a historical context. In this way you can explain the technical advance and emphasise the 'break-though' nature of the development.

Your first listing of ideas from the brain pattern might be as follows:

Mid 1800's — hot filaments
    1879   — Swan, Edison
    1905   — carbon filament bulbs
Metal filaments—
    osmium, tantalum
    1909 — tungsten   —high temperature
Gas filling — nitrogen
           — reduced evaporation
           — but lowered temperature
Coiled filament 1918
           — coiled coil 1934
           — increased temperature
Tungsten halogen
           — iodine
           — high temperature

— long life
— small size
— high light output.

Your next task is to decide on your conclusion. Your basic message is that the light bulb has been developed over a period of more than a century, and yet in recent years a means has been found of producing even more light from a bulb, i.e. the tungsten halogen bulb. The concept of more-light suggests 'Let there be light—and more light'. This ending could link back to an opening 'Let there be light'. Alternatively, the association of light and high temperature, which runs through the speech, suggests the Sun—our source of natural light. The ending could suggest the idea of further possibilities. A check on the surface temperature of the Sun in a suitable reference book would no doubt yield an interesting observation that could provide an arresting opening. It may be better on the other hand to suggest that we might be approaching a limit in light production from hot sources. What about 'cold' light—is this the light of the future? A suitable opening might then make reference to glow worms or to luminous paint.

Clearly there are a number of ways of proceeding. A possible revised set of notes is given below followed by what you might actually say in your presentation

Light Bulbs.
Beginning   — God said
                — development e.l.b. not so simple
Hot bodies   — earlier times
                — electricity
Mid 1800's   — many
                — Edison, Swan 1879
                — carbon—evaporation
                — 1905
Key—higher temp.
                — metal—higher m.p.
                — osmium, tantalum
                — 1909 tungsten

Gas filling	— nitrogen 1913
	— cooling
Coiling 1918	— coiled coil 1934
Limit	— 40 years — breakthrough
	— tungsten halogen
Halogen	— iodine—tungsten iodide
	— 2000 °C—deposits
	— process continues
	— not lost by evaporation
	— 2000 h
Gives	— long life
	— high light output
	— constant light output
Small	— 250 °C
	— optical control
	— spotlights, car
Story	— century
	— end of road?
	— tungsten supreme
	— leapt forward—halogen
	— not higher temp.—'cold'
Policy	— Let there be—and more

Ladies and Gentlemen.

In the beginning, God said 'Let there be light'. And there was light. The development of the electric light bulb was not so simple.

Hot bodies emit light. This had been known from earliest times. With the discovery of electricity came the possibility of producing heat and light from a flowing electric current.

In the mid eighteen hundreds there were many experiments. Eventually, in eighteen seventy nine Edison in America and Swan in England produced electric light bulbs. The filaments were carbon. Threads of organic material were used. These were carbonised by heat. A problem was the limited life of the filament because of evaporation. In spite of this, carbon filament bulbs were in common use by nineteen hundred and five.

The key to more light was higher temperatures. Metal filaments were introduced. High-melting-point metals gave improvements. Osmium and tantalum were tried. Then in nineteen hundred and nine we had the tungsten filament bulb.

To reduce evaporation of the filament, filling the glass bulb with gas was tried. The nitrogen-filled bulb was perfected in nineteen thirteen. Although the evaporation was reduced, the gas caused cooling of the filament and reduced the light output.

The move to higher temperatures continued. Coiling the filament was achieved by nineteen eighteen and by nineteen thirty four we had the coiled coil.

This seemed to be the limit. No further significant improvements took place for forty years or so. Then a breakthrough—the tungsten halogen bulb.

A halogen such as iodine is introduced into the bulb with the gas filling. When the tungsten evaporates from the filament it combines with the iodine. This forms tungsten iodide. The tungsten is now unable to deposit on the glass and blacken it. Furthermore, the tungsten iodide continues to circulate within the bulb. As it approaches the filament, which is at a temperature of over two thousand degrees Centigrade, it breaks down and deposits tungsten back on the filament. The iodine is released and the process continues. In this way the tungsten is recycled and not lost by evaporation. A life of two thousand hours can readily be achieved.

Thus the bulb gives long life, high light output and constant light output throughout its life. It is made small in size because the chemical reaction needs a temperature of at least two hundred and fifty degrees Centigrade. Its small size allows precise optical control. For this reason the bulb has become popular for shop spotlights and car headlights.

Our story has covered more than a century. Are we now at the end of the road? Probably not. The tungsten filament bulb was supreme for forty years. But then, technology leapt forward and gave us something even better—the tungsten halogen bulb. No doubt there will be further advances. Perhaps it will not be in the attainment of higher temperatures that the next important step will be taken. It may be 'cold' light that we shall see developed. Whatever the means, the policy will be 'Let there be light—and more light'.

The speech contains 510 words and therefore fits properly into a five minute presentation. There are 48 statements giving an average of about 11 words per statement.

6.4 IMPROMPTU SPEAKING

The situation we dread most of all is impromptu speaking or speaking 'off the cuff' as we refer to it. A teacher I had would often say how wonderful the human brain is. It starts working before we are born, it continues, waking or sleeping, until the day we die and the only time it stops is when someone asks us to stand up and say a few words.

Again it is necessary to emphasise the value of practice in developing a skill for speaking without preparation. Chapter 14 gives methods of achieving improvement in the long term but application of the following technique will give some immediate improvement.

Avoid starting hurriedly. Take two slow fairly deep breaths as recommended for all presentations. This is particularly important if you are replying to criticism, provocation or an awkward question. The pause before starting will calm the situation, will calm your own tenseness and will allow you to think what your opening statement is to be.

Use your opening to set the scene. Define the subject, review the position or restate the question but do not commit yourself to an answer or a point of view. What you are doing is creating a little thinking time. In addition you are testing your voice and the resonance of the room and you are preventing the possibility of you 'drying up'.

During this thinking time you must decide on your conclusion. This is to be the essence of your message, the answer to the question, the package of information that you wish the audience to retain. Having fixed the idea of your conclusion in your mind you should consider it to be not only the ending but also an escape route. If you run into any kind of difficulty from this moment on you can go straight to your conclusion. In fact, once you know that you have an escape route you will feel quite confident and will probably not need to use it as such.

With your ending or escape route clearly established you can present the detailed points you wish to make. Be concise. Avoid repeating statements simply in order to improve the wording. Repetition is permissible but it must be done for effect, not by way of correction or improvement. If you have, say, four points you wish to make, you will be tempted to start by saying 'firstly' or 'one' or 'A', intending to work your way through numerically or alphabetically. It is better not to do this. One danger is that you will transfer to a different listing system and produce something like 'A, . . . two, . . . and thirdly . . . .' Worse still you will forget one or more of your points and having promised four you will present no more than three. It is better to simply start with your first point, then move to the next one, and so on. This leaves you free to omit points if necessary, either because you have had second thoughts about their usefulness or perhaps because you have a time problem.

Use short sentences with simple constructions: lengthy subordinate clauses will result in you forgetting what the subject of the main verb was. Rather than saying 'The experiment which was carried out three times on each of five consecutive days yielded results which agreed within three per cent,' it is better to say 'We carried out the experiment fifteen times—three times a day for five days. The results agreed within three per cent.'

Note the change in style in the above example. Keep the style conversational. Attempting to give a formal air will increase the probability of tangled sentences and the listeners will be paying more attention to your errors of grammar and syntax than your message.

As you present your conclusion, signal that it is the end by a slower pace and by a descending tone. Speak with an air of authority and finality. Imagine you are a judge sentencing a criminal to imprisonment for life. Pause after the last word, then sit down. Resist the temptation to add a further comment. Do not say 'thank you' and do not apologise for anything.

7   Visual Aids

7.1   GENERAL

Visual aids are not a show in themselves: they are, as the name suggests, to aid the speaker in making his presentation more effective. The common example of misuse of visual aids is given by the speaker who puts a selection of slides in his bag and heads for the lecture hall. The talk consists of 'the next slide shows . . . the next slide shows . . ..' And then we get 'the next slide shows . . . oh, no, sorry, this is another example of what I was talking about earlier . . .' or 'what is the next slide? Oh, yes, this is a . . .', or even 'this is a slide I just slipped in. It isn't very interesting . . ..'

Visual aids should be used when they are necessary or useful to the speaker in getting his message across to his audience. They can save words by providing or clarifying explanations. They can illustrate relationships and consolidate information, and they can attract and hold attention. In holding attention they can of course distract the audience from what the speaker is saying and this must be appreciated.

Choose your aids carefully at the time of preparing your speech. Be clear as to what it is you are attempting to explain or illustrate. 'Every picture tells a story' but it must be the right story. Pictures should not be used simply because they are readily available. A picture should be used because it is right for the purpose. It is no use saying 'I am afraid this isn't a very good slide but . . . .' If it isn't a good slide why use it?

The various aids available[5] are described in the following sections and a choice can be made on the basis of the advantages and disadvantages of each in relation to the particular requirement. The choice can be limited at the outset, however, because of the size of the audience. Many visual aids cannot be used for large audiences because the viewing distance from the back row would be too great. The question of legibility of lettering viewed at a distance is discussed in detail in §7.9 where instructions for the preparation of slides are given. It is, however, appropriate to give here some general guidance on the relation between audience size and viewing distance. In planning your presentation you will know the approximate size of the audience expected, but you may have no idea of the size of room in which the meeting is to take place.

If the meeting has been arranged on the basis that visual aids will or may be used then you must assume that the seating arrangement has been planned accordingly. Your task is to ensure that your aids are satisfactory given a satisfactory seating arrangement. In other words it would be wrong to assume that the organisers will be using, for example, an overcrowded small room. This might bring the back row closer to the speaker and improve the visibility of your aids but it would make satisfactory viewing impossible for many in the audience and difficult for all.

Seating capacity is discussed in detail in Seekings.[6] From data given by the author, figure 7 has been constructed. This shows, for three values of room length to width ratio, the length of room required to accommodate an audience of given size. Two scales of seating capacity are included: the first relates to a theatre arrangement, i.e. rows of seats, while the second, classroom arrangement, allows for each person to be

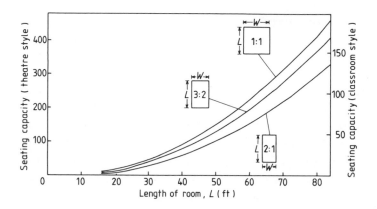

Figure 7   Seating capacity of rooms.

seated at a table. The values of capacity shown assume that the arrangement is suitable for the presentation of slides. If visual aids were not to be used the seating capacity would of course be much greater than the value predicted from figure 7. This is because there would be no minimum distance to the front row to consider and no limitation imposed by the viewing angle for people sitting off-centre.

With a knowledge of the approximate size of audience it is thus possible from figure 7 to estimate the maximum viewing distance. It may be necessary to guess the room shape, but the likely error involved can be appreciated from the graphs. The suitability or otherwise of visual aids you are considering using can be judged in relation to the predicted value of maximum viewing distance.

A final cautionary note will bring to an end these introductory remarks relating to the use of visual aids. Regardless of which aids you choose you must ensure that the mechanics of operating the aids will create no problems. Everything must be verified as being in working order and you must know where the controls are and how to operate them. The onus is on you as the speaker to satisfy yourself in advance that everything is as it should be. It is extremely discourteous to members of the audience to waste time by subjecting them to a demonstration of the range of focusing capability of a projector or a search for a piece of chalk.

7.2 THE CHALKBOARD

That search for the piece of chalk brings us to the chalkboard. This is the most common of visual aids and has therefore the great advantage of being readily available. It also has the advantage that there is very little to go wrong. Chalk should be provided of course, but why not carry a piece with you?

With regard to preparation make sure that the board is cleaned before you start, preferably with a damp cloth. Check the position of the board relative to the lighting and move the board if necessary to avoid glare. Note the size and shape of the board and decide exactly where your words or diagrams are to fit. Prepare a model layout on one of your note cards. This will ensure that you do not run out of space. Erasing is possible of course, but this should be planned erasing and not casual. Distinguish between items that you wish to retain for later reference, items which should be removed once dealt with, and items which should be retained for your ending.

When writing on the board use block capitals sparingly: lower-case printing is more restful to the audience. Handwriting can be difficult to read. Walk along as you chalk so that your lines of lettering do not droop at the end. Above all, do not speak as you are writing. Even if the room is small and you can be heard, you will be holding very little attention while you are facing the board. Be on the lookout for spelling mistakes which can arise very easily when chalking on a board.

A chalkboard can be used very effectively to build up a summary of your talk by means of headings and simple diagrams. Coloured chalk can be used for emphasis.

A disadvantage of the chalkboard is that time is used in writing or drawing. There is also a limit to the amount of material that you can present without erasing, though this is rarely a serious handicap. More important is the limit on the room size. A chalkboard is not effective in a large hall because of the limited line thickness produced by the chalk.

7.3   THE WHITE BOARD

A variant of the chalkboard is the white board, a sheet of plastic on which felt-tip pens are used. The technique of use is similar, and much of what has been said in the preceding paragraphs is equally applicable here.

Pens containing water-based ink are commonly used and erasing is then by means of a damp cloth. A useful technique, however, is to use pens with spirit-based ink for parts of the presentation. Such parts will not be erased by the damp cloth so selective erasing and modification of diagrams, for example, is possible. Methylated spirit or a proprietary fluid is used to clean off the spirit-based ink.

The presentation is visually more attractive than it is possible to achieve with a chalkboard, the colours being more striking. Chalk dust on the hands and clothes is avoided and sensitive members of the audience appreciate the absence of 'screeching' chalk!

However, there is slightly more that can go wrong than with a chalk board. Felt-tip pens run out without warning. Make sure there are spares of each colour you require. If you are using spirit-based and water-based pens be careful that you do not confuse the two types. It is safer, if possible, to prepare the spirit-based part of your presentation beforehand and use only the water-based ink pens during the presentation.

7.4   THE FLIP CHART

A flip chart is rather like a giant writing pad. Measuring about 3ft by 2ft it contains sheets of plain white paper. A large clip is used to mount the pad on a board. A common arrangement is an easel carrying a chalkboard, but purpose-built portable flip-chart frames are available. Felt-tip pens are used to write on the sheets. You can use the sheets as you would a chalkboard, limiting the amount of material on any one sheet and turning the used sheets as you progress so that they hang over the back of the board. Alternatively, you can tear off each sheet as you finish with it, screw it up and drop it in the waste bin. This technique can be very effective.

An advantage over the chalkboard is that you can lightly pencil in your sheets before you start. This ensures that everything is spaced well, as you simply go over the pencil lines with your felt-tip pen. The procedure also ensures that nothing is overlooked. You can of course prepare the sheets fully beforehand and simply turn them over as required. If you do this, remember to leave a blank sheet below each prepared sheet. Otherwise the writing and diagrams will show through faintly on the sheet above. The full potential of the flip chart can be achieved by preparing part of the presentation beforehand. The rest, to be developed as the talk proceeds, is pencilled in.

If you provide your own flip chart, ready prepared, and your own pens then there is little that can go wrong. You must of course ensure that there is available a suitable means of mounting the chart, and do not forget to take spare pens!

7.5   THE MAGNETIC BOARD

A steel sheet provides a surface on which shapes backed with magnets can be mounted. Magnetic plastic strips, sheets and string can be obtained for use with such boards, in a variety of colours. Letters and numerals are also available.

A magnetic board is particularly useful when movement of shapes from one position to another is needed. The operation of a mechanical device, for example, can be demonstrated much more readily than by means of a diagram. Displays can be added to as an argument is developed and this can not only make the information easier to absorb but can effectively hold the interest of the audience.

Some magnetic boards are painted so that they can be used as chalkboards. Others are overlaid with plastic and can be used as white boards.

7.6   THE FELT BOARD

The felt board consists of a rigid board covered with felt or similar material. The words or diagrams are drawn on special thin card which is available in various colours. Flock is attached to the back of the card and this causes the card to stick to the felt when pressed against it.

By using a variety of shapes and colours it is possible to build up a very eye-catching presentation. The lettering or drawings are best done using a broad black felt-tip pen to give the necessary contrast on the various colours.

Remember to press the cards firmly against the felt when positioning them. The adhesion is less than on a magnetic board and furthermore it is not possible to slide the cards about. For this reason it is not advisable to use a felt board when movement of the parts of the display is required. A magnetic board should be used in such circumstances.

Felt boards are not always available but improvisation is possible. You can carry with you a piece of felt, folded up in your briefcase, and you can drape this over a chalkboard and easel. If you have difficulty obtaining the special card you can use ordinary coloured card. The fixing is then by means of small pieces of Velcro (a non-sticky tape designed to grip fabric) attached to the corners.

You will find a felt board useful when the time for your presentation is very limited. You waste no time in writing and yet the visual display develops gradually as on a chalkboard.

7.7   THE OVERHEAD PROJECTOR

None of the visual aids described so far are suitable for use in large halls. It is necessary to use a projector of some kind to give sufficient magnification when members of the audience are seated a long way from the speaker.

The overhead projector[7] has become very popular in recent years and is now usually provided at technical conferences as a matter of course. It consists essentially of a box, the top surface of which carries a glass window approximately 25 cm square. A bulb in the box illuminates the window from below. The material to be projected is prepared on a transparency which is laid on the window. Light passing upwards through the transparency enters the focusing head which is mounted on a vertical arm rising from the light box. The focusing head projects the image onto a screen mounted on the wall behind the speaker.

The speaker stands facing the audience with the projector in front of him, but slightly to one side. The image is projected past his shoulder to the screen behind and above him. The speaker thus has no need to turn and face the screen: he has sight and control of the transparency in front of him on the window of the projector. On a vertical screen the image from the projector is distorted, being wide at the top and tapering to the bottom. To avoid this distraction it is usual to lean the screen forward at the top.

Transparencies are prepared from transparent plastic film. Felt-tip pens, of the water-soluble type, can be used for lettering and drawing and the projector can therefore be used as a chalkboard, the speaker writing as his talk proceeds. The projector can be fitted with a roll of plastic film which can be wound across the window as required.

More permanent transparencies can be prepared using a copying machine but best of all are those produced photographically. Transparencies can thus be an assortment of photographs, diagrams and lettering. Typescript can be copied to make a transparency, but it is rarely sufficiently legible and should be avoided, unless the room in which the presentation is to be made is small. Even then a good typewriter with a golfball head should be used. Cardboard mounts ensure that the transparencies do not curl when placed on the projector. The mount also provides space on which notes can be written.

The correct way to use the projector is to place the transparency on the window, then switch on the light. When changing the transparency, switch off the light first and, when the next transparency is in position, switch it on again. In this way you do not subject the audience to the glare of a fully illuminated screen. Obviously you need to know where the on-off switch is. Find out beforehand. The audience have no wish to see you experimenting with the apparatus. The same applies to the focusing arrangement. Make sure you know how to alter it if necessary. You need to give thought to your plan of action in the event of something going wrong. What do you do if the bulb blows or the power fails? Question the organiser or chairman before the meeting starts. Is there someone available to deal with a breakdown or will you be expected to carry on as best you can without the projector?

Remember when using the projector to stand to one side otherwise you will block part of the projected beam. It is not difficult to remember this when you are merely presenting prepared transparencies, but if you are writing or drawing you will be tempted to stand directly behind the projector. Writing or drawing prompts a further warning. The shadow of your hand will be in view as you write and you may be trembling slightly! Your apparently straight line may appear magnified on the screen as a sine wave and any physicists in the audience will spend the next few minutes estimating the frequency of your tremble!

If you need to point out some feature on your transparency, you can do this without turning to the screen. Take a pencil and lay it on the transparency with the point aimed at the item of interest. Holding the pencil while pointing is not advisable, not only because of the risk of tremble but because the outline of the pencil will be out of focus and will produce a diffuse shadow on the screen. Even though you will not be tempted to stare at the screen while speaking, remember that neither should you stare at your transparencies as they rest on the projector window. The place for your eyes is on the audience except when you need to glance at your notes, or manipulate your transparencies.

A particularly useful technique that the overhead projector allows is the building up of complex diagrams in stages. The transparency is built up in a series of layers, each hinged so that it falls into place on the one below. Each layer carries a part of the diagram and different colours can be used as required. Electrical circuits, maps or plans can be explained clearly by means of a layered transparency. It is of course possible to return to an earlier stage in order to recap or bring in a further consideration, whereas a chalkboard or flip chart does not allow this.

Opaque material can be used for layers if it is desired to mask particular areas. With a little ingenuity it is possible to construct transparencies with sliding or rotating layers to give movement in the diagrams.

Note that an overhead projector will cast a silhouette of an object placed on the window. This facility can sometimes be used to good effect. Exhibits with distinctive shapes can be shown: for example, small gears or other mechanical devices can be illustrated with a touch of realism. On one occasion interest in a talk on metal fatigue was revived by projecting the silhouette of a paper clip!

7.8   THE SLIDE PROJECTOR

The slide projector is long established and perhaps still the most common visual aid used for presentations to all but small audiences. Traditionally, the projector is mounted at the back of the room or hall and operated by a projectionist. Remotely operated projectors are now becoming widely used, the speaker having control from a hand-held unit.

Slides are prepared from 35 mm film, coloured or black and white, and are mounted in plastic frames to give an overall size of 50 mm square. Slides tend to have a greater permanency than any of the other forms of visual material so far described.

It is all the more important therefore that they are prepared in a satisfactory manner. A later part of this chapter is devoted to the preparation of slides; the remainder of this section will deal with the speaker's use of slides in his presentation.

Because of the permanency of slides you will often find that you are choosing slides for use rather than preparing them. Many organisations retain a library of slides from which speakers can make a selection suitable for the occasion. The rule to follow in making your selection is that the slide must be an aid to what you are trying to say. Your message is primary, the slide is secondary. You cannot select your slides until you know what your message is. Your talk must not be a commentary on slides that just happen to be available. If the slide does not illustrate your point, then you must not use it. You cannot waste valuable time explaining why the slide does not show what you hoped it might show or what it should have shown. If a slide is not quite right then prepare one that is.

Having selected your slides you need to number them in sequence. Small numbered circular self-adhesive labels can be purchased for this purpose. They are easily removed so the numbers can readily be changed if the slides are later used in a different talk. If you locate the numbers in a consistent position on each slide you will ensure that all slides are the correct way up and the correct way round when they come to be used. The bottom left-hand corner when the slide is held for hand viewing is the usual position for marking.

If you are fortunate enough to be provided with a projectionist, make contact with him well before the meeting starts. Make sure he understands how you have coded the slides with regard to top and bottom and back and front. Explain to him your plan of presentation, that is to say whether the slides will be shown in sequence with no breaks or whether you will request at intervals the projector to be switched off and the hall lights switched on. Be especially careful if you intend to return to an earlier slide. The projectionist loads the next slide as the previous one is shown, so an unexpected call for 'the third previous slide' can cause an embarrassing delay. It is far better to have two copies of the same slide prepared and avoid backtracking completely. Suggest to the projectionist that you will call for each slide verbally. Signals involving tapping with the pointer or stamping the foot can lead to misunderstanding and often appear comical.

If you are using a remotely controlled projector load the slide magazine yourself beforehand, checking that you have each slide the correct way up and the correct way round. Check the remote-control unit. This allows you to alter the focus and to advance the slides or backtrack. The slide controls are usually marked with no more than arrows and it is easy to confuse the forward and backward buttons. Stick a piece of coloured paper on the forward button and avoid any reviewing of slides if possible. You will not be able to switch the projector on or off remotely. This means that you will need a number of blank slides that project black. Insert a blank slide at the start, at the end, and at any other required position to give you a blackened screen. In this way you can switch the projector on before you start, and leave it running throughout your talk. The distracting 'Could someone please switch the projector off' is thus avoided.

Experiment before the meeting starts to establish whether you will need to darken the room while the slides are being shown. It is preferable to avoid repeated lightening and darkening of the room during your talk if possible. This is especially so if you are not able to control the lights from where you will be standing.

As was mentioned when discussing the overhead projector, you need to consider what may go wrong. If the projector breaks down, is there someone on hand to deal with repairs? Is there a spare projector available? Will you be expected to carry on without your slides? The time to think about such questions is before the meeting starts, not when the emergency has arisen.

When presenting your slides avoid talking to the screen. You must talk to the audience. The screen of course is behind you, probably to one side and until you look at the screen you have no way of knowing whether the slide is the correct one or whether it is the right way up. As each slide appears, turn and look at it but remain silent. Then turn back to the audience and continue speaking. I have seen a rather ingenious idea which avoids the need to turn to the screen. A car driving mirror (preferably convex) mounted on a stand can be positioned on the lectern or on a table near the speaker to give a view of the screen.

Problems can arise if you wish to point out various features on the slides. Pointers are a nuisance. If you must use one, force yourself to accept a strict discipline. Pick up the pointer, point and put down the pointer. Do not speak while pointing. Failure to keep to these rules can spoil your presentation. You will find yourself repeatedly pointing at the same feature, speaking to the screen and performing various feats with the pointer, some more worthy of a circus act. I have of course been speaking of the old-fashioned pointer, a stick in fact. There are now torch pointers which project an illuminated arrow on to the screen. These are technically more interesting but I have never seen one used well. The arrow darts violently across the hall before homing in on the screen and after overshooting several times finally comes to rest on its target. I say 'comes to rest' but often it continues to oscillate to and fro magnifying the nervous tremble of the speaker's hand.

It is better to avoid pointing if you possibly can. Try using words to explain where your audience should look. 'In the bottom left-hand corner we can see . . .' 'Right in the centre of the slide is the . . .' 'If we look at the top half of the slide first, we can see . . ..' When preparing your slides have in mind the parts you will draw particular attention to. You can insert arrows, stars or other symbols on the slides to pick out certain features. Prepare your slides not as isolated visual displays but relate them to what you wish to say.

7.9   PREPARING MATERIAL FOR SLIDES

All visual material, whether it be for projection from a slide or for producing on a chalkboard, should conform to a number of basic rules. The rules are designed for the prevention of cruelty to audiences. The material must be clear and comfortable to view and the lettering must be legible. If the presentation can be made attractive and interesting then this is a useful bonus.

The points now to be covered apply therefore to all kinds of visual aids. However, the discussion will be mainly in relation to slides or transparencies. This is because there is a natural tendency when writing or drawing on a chalkboard, say, to get everything about right in terms of size. When preparing slide material there is no such natural feel for the appearance of the magnified image on the screen.

The resolving power of the eye is usually taken to be one minute of arc. Expressing this as a ratio of the distance at which we can resolve an object to the size of the object, we get 3500 : 1 approximately. This means that the thickness of a line used in a diagram or lettering must at least meet this criterion for visibility. If we form a series of numerals or letters of the alphabet we will have a height of lettering at least five times the line thickness. This is evident if we consider the maximum number of horizontal intercepts that are possible taking a vertical section through each numeral or letter. '2', '3', '8', 'B', 'E', 'S', for example, have three such intercepts, the maximum possible. Allowing for two white spaces to separate the three black lines gives a total of five zones, so the height must be at least five times the line thickness. Our ratio expressed in terms of viewing distance to lettering height is therefore 700 : 1.

Although opticians may use eyesight test cards with lettering based on a ratio similar in value to this, for purposes of visual aids further factors have to be taken into account. Most styles of lettering use a height of letter which is greater than the minimum value of five times the line width. The height may be up to ten times the line width. This brings the ratio of viewing distance to height down to about 400 : 1 and this can be taken as the minimum practical limit for visibility. For comfortable viewing a further factor of two should be allowed so that the rule to work to is that the height of lettering should be not less than 1/200 times the viewing distance. Thus, at a distance of 5 m lettering should be at least 25 mm high. It is worth noting for purposes of ready comparison that the moon's diameter is about 1/100 times its distance from the earth.

It is also worth noting that licensed drivers of motor vehicles are required to be able to read a car number plate at a distance of 25 yards. The height of lettering ($3\frac{1}{2}$ inch) is such that it represents 1/260 times the viewing distance. The style of lettering used on number plates has a letter height just over five times the line thickness to give maximum legibility, so the legal requirements are fairly generous. They do however represent approximately the degree of legibility to be aimed for in preparing visual aids.

The suitable size of lettering for slide production depends on the layout size and to some extent on variations in screen size. The British Standards Institution recommends that an A4 sheet (297 mm × 210 mm) be used for the layout.[8] Within this area a neat area of 247 mm width × 165 mm height is defined. This is the maximum possible area that can be shown on the screen. A smaller area, the safe area (225 mm × 150 mm), is the maximum area certain to be shown on the screen allowing for variations in mounting and projection. The sizes of letters or symbols suitable for such a layout range from 6 mm for main titles to 4 mm minimum. Recommended line thickness varies from 1.5 mm (thick) to 0.4 mm (thin). It is suggested that the legibility of the finished layout be checked by viewing it at a distance of six times its width.

An American journal[9] gives guidance on letter sizes assuming a layout size of 9 in × 6 in. The preferred letter size is 5/16 in and the minimum 5/32 in. This minimum is similar to the British Standards recommended minimum.

If the above recommendations are interpreted in terms of a transparency for an overhead projector, the minimum letter size is found to be about 5 mm.

In figure 8 the recommendations relating to lettering size in

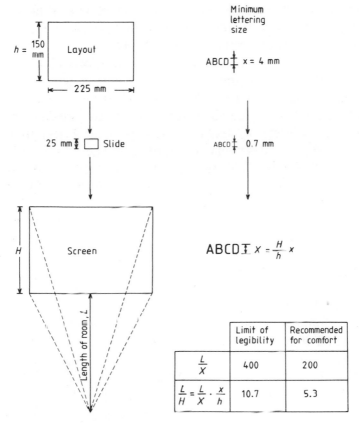

Figure 8 Legibility of lettering on slides.

slide preparation are summarised diagramatically. It can be seen that the legibility of the lettering depends on the height of the projected slide image on the screen and the length of the room. The speaker usually has no control over these factors and can but assume that the organisers have followed recommended practice. In Powell[5] it is recommended that the ratio of room length to image height (L/H) should be not greater than eight. In figure 8 the rule of 1:200 for comfortable viewing is seen to lead to a value of L/H of 5.3. The 1:400 limit of legibility leads to a value of L/H of 10.7. The recommended value of eight lies midway between these two values and appears therefore to have been based on a lettering height to viewing distance ratio of 1 : 300.

It is important not to crowd too much information on one slide. A lengthy paragraph of text will interfere with audience interest. Some will read the text and ignore the speaker. Others will ignore the slide and listen to the speaker. Most however, will simply feel confused. It is far better to have several simple slides than one complicated one. If a list is to be presented, for example, prepare one slide with the first item only shown. The second slide should include the first two items, and so on. In this way each item is added as it is mentioned and the audience needs to absorb only one idea at a time. The temptation to read ahead of the speaker is usefully removed.

In written information no more than seven or eight words to a line or seven or eight lines should be used on one slide. In tabulated data three columns of numbers and eight lines, or five columns and five lines, are ample. Graphs should be prepared with broad lines and only a few grid lines should be shown. Titles should be bold and short.

Graphs or diagrams prepared for reports or publications are not suitable for use as slide material. The lettering will be too small, the lines too thin and the layout too crowded. Even worse are detailed engineering drawings. Typescript is not usually suitable, though many speakers insist on showing large typed tables of figures probably extracted from a report. A more recent trend, and equally unsuitable, is the use of computer printout material. Typescript may be acceptable if it is produced on a good typewriter with a golfball head and if plenty of space with the minimum of words is allowed.

Slides showing diagrams or text can be produced as negatives, i.e. with white lines and lettering on a black background. This is known as the reversed text technique. Not only is the processing cheaper but the slides are improved in a number of ways. Legibility is improved, viewing is more comfortable and dust and dirt on the slide is less obvious. In addition, viewing is possible in a partially lit room. Colour can easily be introduced to give alternatives to white lettering.

7.10 OTHER POSSIBILITIES

It is not possible to differentiate completely between a public speaking aid and a teaching aid. All of the aids so far described are teaching aids and most of the guides to their characteristics and use are written from that point of view. They are also extremely useful public speaking aids as the foregoing sections have described.

However, not all teaching aids are public speaking aids. Films with soundtrack, for example, may be an excellent means of informing an audience, but are a substitute for, rather than an aid to, public speaking. A film without a soundtrack may or may not be a public speaking aid depending on how it is used. The point to note is that in public speaking the speaker controls the flow of information to the audience. In a sense all visual aids may remove a degree of control from the speaker. A slide, for example, showing a view of a piece of farming machinery may also show a little of the surrounding countryside. This may be harmless enough in reality but to illustrate the argument it can be supposed that some member of the audience will be distracted. He may think he recognises the location as the village he spent his summer holidays in. In wrestling with this feeling of *déjà vu* he may well miss the speaker's comment on the slide. As the information communicated by aids becomes more complex, so the speaker's control of the information flow becomes less. A working model of a machine is an excellent teaching aid but many different pieces of information are transmitted from it to the observer simultaneously. The observer will select by concentrating on the various bits, not necessarily in a logical order. If such a model is used as a public speaking aid there is a danger of audience distraction and loss of control by the speaker. It is very noticeable that during demonstrations to groups of people the start of the demonstration acts like a signal for people to start murmuring among themselves and to ignore the speaker. It is as if the audience recognises the model as the new, albeit inanimate, communicator.

These general remarks will allow you to make decisions about various kinds of aids that you may be considering using. It would be quite wrong to be dogmatic about which aids should or should not be used. Novelty is to be encouraged. As mentioned many times the role of the speaker is to attract and hold attention. He should therefore be encouraged to explore new ways of doing this. If he wishes to pedal round the platform on a unicycle, who is to say this is wrong?

Films, closed-circuit television, working models and demonstrations can all be used effectively but you must be aware of the danger of distraction and loss of control. Reduce the possibility of distraction by removing as much unwanted information as possible. Film or television presentations should be exactly right for the talk you are giving and not used merely because they are available. A working model may give too much information or raise too many questions. Consider whether an idealised moveable model on a magnetic board or an overhead projector transparency with overlays might not be more suitable.

As has been mentioned several times previously, make sure that your apparatus works satisfactorily. As we all remember from our student days, demonstrations have a habit of going wrong, though they may have been satisfactory in rehearsal. It is impossible, of course, to ensure with 100% certainty that a demonstration will go according to plan, but one thing you can and must do is decide on your escape route in advance. Decide on what you will do and what you will say if your demonstration runs into problems.

Part II

Ways of Improving

8 Confidence

Lack of confidence has its origins in the fear of public speaking. It is important to recognise that this fear is perfectly natural and, indeed, desirable. The aim in developing confidence is to control fear, not to overcome it. Do not imagine that the day will come when you are not nervous of speaking in public. Rather, look forward to the day when you will be confident enough to undertake any public speaking commitment, knowing that you will make a good job of it but knowing also that you will be nervous when the time comes.

Some speakers will assure you that they are not nervous. Treat this with suspicion. I hold the opinion that there is only one type of speaker who is not nervous. He is the man who has solved his fear problem by isolating himself completely from his audience. In effect he talks to himself rather than to his audience. He does little or no preparation because he 'knows his subject'. He overruns his time and is completely oblivious of the extent to which he is boring his audience or keeping other speakers waiting. I say 'boring' because this is usually what happens. He has no sensitivity to his listeners, he is not really aware of their presence. He is fluent, listens to what he says and judges it to be of good quality. Unfortunately, such speakers can never improve. They have insulated themselves from feedback and never see any need for instruction or tuition. This is not the kind of speaker you wish to become. The effective speaker is nervous but he knows that his nervousness is under control. He recognises that his body needs the stimulus arising from fear in order to give of its best. The body produces adrenalin in such circumstances to heigh-

73

ten its performance. A greater sensitivity to the audience, and to the situation in general, results in a more effective communication between the speaker and the audience. It is not just public speakers who experience such sensations. Many great actors and actresses admit to being extremely nervous before performances.

The aim, then, is to control the fear. How is this done? There are two parts to the process; the first involves adopting a particular state of mind, and the second involves practising a few exercises. The correct state of mind comes from acceptance of the following propositions. Read them frequently, particularly before any occasion when you are to speak in public. They are written in the first person to emphasise the need for you to believe them deeply rather than to merely give them passing attention.

(1) I have established the correct procedures to adopt in making this speech. I know more about the subject of public speaking than do the members of the audience and what I am about to do will be done correctly.

(2) I have prepared my material well and have used the proper techniques. The chance of me making any mistakes is therefore small.

(3) Even if I make a mistake it will be much less noticeable to the audience than it is to me.

(4) If I run into difficulties I have my escape routes. I can resort to reading the full text of my prepared speech, or in the case of an impromptu speech I can move directly to my ending. If my mind should go completely blank, I can take out my handkerchief and blow my nose, pour myself a glass of water or ask if someone would mind opening a window.

(5) I am undertaking this engagement because I have something worth communicating to the audience. With respect to this particular message I am the expert. Furthermore, the audience see me as the expert and are prepared to accept me as such. Their attitude to me is friendly even if they disagree with what I say. The last thing they want to happen is

that I do not succeed. Any embarrassment I suffer, they too will suffer.

It can be seen that the principle being used is that confidence comes essentially from knowledge. Knowledge of the message to be communicated, together with knowledge of the technique of doing it, is the key. From the knowledge comes the awareness of having the knowledge. This step may seem obvious but is too easily overlooked. Awareness is a subconscious acceptance but may need repeated conscious concentration before developing. For this reason the five points listed above have been constructed with a flavour of indoctrination.

Having adopted the correct state of mind you need to master three techniques concerned with the physical side of control of nervousness.

The physical symptoms of nervousness are well known: heart beating faster, butterflies in the stomach, dry mouth, desire to swallow, perspiration and so on. Such symptoms are perfectly natural. Accept them as such, welcome them. This will prevent panic setting in. When you start your speech you bring these symptoms under control by three means. The first is correct stance, the second is correct breathing and the third is speaking slowly.

Stance is important because you must look right and feel right. Unless you feel right you will not feel confident. A suitable stance to adopt is achieved in the following way. Stand with the feet slightly apart and with one foot slightly in front of the other. This gives a diagonal support to the body and minimises any tendency to sway from side to side or to rock forward and back. The best place for the hands is hanging limp at the side of the body, though if notes are to be used one hand will be occupied in holding the cards well up in front of the body. The body should be relaxed to the extent that the arms can be felt to oscillate slightly as pendulums under their own weight.

The stance as described will look right. The problem is that it will not, at first, feel right. You will imagine that your arms are about four feet long and that your hands are the size of dinner plates. You will be tempted to hide your hands behind your back or in your pockets or to fold your arms and tuck your fingers down the inside of your jacket sleeves. Such antics will certainly not look right. There is a difficulty in accepting that the stance gives you a perfectly acceptable appearance. You look normal, relaxed and confident. You have to come to believe that this is so, and, when you do, you will find that as you adopt your stance ready to start your speech you will feel relaxed and confident. It takes a little time and practice to get used to the feeling of the stance and to develop the necessary faith in its appropriateness. For this reason the next chapter takes the subject of stance further, and provides some simple exercises designed to speed up the process of developing a correct stance.

Correct breathing is the second way in which the symptoms of nervousness are brought under control. Again a later chapter, Chapter 10, will discuss breathing in more detail and provide u ful exercises. It will suffice here to say that slow breathing is essential. Develop the habit of always taking two slow breaths before starting your speech: thus, in . . . out . . . in . . . and speak. The tendency to breathe quickly is associated with shallow breathing and it is not easy to overcome. A little time spent in breathing exercises is worthwhile because good breathing is necessary not only to give confidence but also to give good speech production as is explained later.

The third and final way of bringing the symptoms of nervousness under control is by speaking slowly. More is said in Chapter 12 about developing a correct speaking pace and appropriate exercises are given. For several reasons you should not speak faster than about 100 words per minute. One reason is the link between nervousness and speed of speaking. Nervousness causes us to speak more quickly but, less obviously, rapid speaking increases our state of tension. Thus, we are in danger of entering a state of positive feedback in which our psychological disposition rapidly gets worse. By speaking slowly we remove tension and calm our nerves.

To summarise, then, confidence comes from knowledge; knowledge of techniques of speaking and knowledge of the material to be presented. Nervousness is natural and to be welcomed. Its symptoms are controlled by adopting a correct stance, correct breathing and by speaking slowly.

A final point to make is that confidence improves with experience. You should therefore take every opportunity of speaking in public. Asking questions is a good way of getting a little practice with a minimum of preparation. If you are scheduled to make a lengthy presentation at a meeting, try to take the opportunity of making a short statement prior to your presentation. Obviously, before you stand up to speak you must have something worth saying.

9 *Stance*

It is necessary to develop the technique of adopting a suitable stance so that it becomes instinctive. In this chapter a number of exercises which will help in this development are described.

First, it is appropriate to recapitulate on what constitutes a suitable stance and why it is necessary to adopt one. As explained in the preceding chapter a suitable stance is necessary to give confidence. If you feel relaxed and natural and know that you look relaxed and natural, then you will feel confident. A suitable stance is also necessary from the point of view of the audience. Your listeners should be concentrating on your message and should not be distracted by any peculiarities in your stance. Any swaying, swinging, standing on one leg or climbing on the furniture will result in a loss of audience attention.

It is always better to speak from a standing position if there is any choice in the matter. Obviously there are occasions when protocol demands that you speak from a sitting position. A sitting position is by its very nature more relaxed and speakers will often choose to sit for this reason. A standing position is better because it encourages better breathing, and therefore greater volume of sound and clearer articulation. It also holds the attention of the audience more effectively.

EXERCISE 1: THE ARM DROP

Stand in front of a full-length mirror with feet slightly apart. Raise the arms above the head. Bend the knees very slightly, just enough to ensure that the legs are relaxed and not rigid. Let the arms relax by bending the elbows and let the hands hang limp on the wrists. Move the weight of the body forward onto the balls of the feet and raise and lower the heels slightly. Create a feeling of increasing relaxation by repeated slight raising and lowering of the heels. Increase and decrease the bend of the elbows several times to give a relaxed feeling in the arms, but maintain the arms in a raised position above the head.

As the arms begin to tire, prepare to let them drop to the sides of the body. Decide that quite suddenly you will cease to hold any control over your arms. Lower the heels so that the weight is distributed over each foot and then let the arms drop. They should drop under their own weight without any control and should swing slightly before coming to rest at the sides of the body. Rotate the upper part of the body, from the waist, to and fro several times to encourage a slight swinging of the arms. This will ensure that the arms are still relaxed. The hands too should still be hanging relaxed.

Now place one foot slightly forward of the other and take a good look at yourself in the mirror. You will see yourself in the stance described in the preceding chapter. Spend some time observing. You will see a person who looks relaxed and natural. Concentrate on how you feel in order to develop an association between how you feel and how you look.

By repeating this exercise over a period of time you will develop an instinctive awareness that when you adopt the stance it does in fact look relaxed and natural. This awareness will lead to a relaxed and natural feeling each time you adopt the stance.

EXERCISE 2: THE FORWARD-AND-BACK SWING

Stand with feet slightly apart and arms hanging at the side. Bend the knees very slightly so that the legs are not rigid. Spring slightly on the feet several times to encourage a relaxed feeling. Now bend forward from the waist and swing the arms backwards. Maintain the swinging rhythm and bring the body upright again, allowing the arms to swing forward of the body. Repeat the forward and backward swing, each time bringing the arms higher in front of the body. The rhythm should be unhurried, and progress with a natural and relaxed feeling. The arms should be extended but not rigid. The hands should hang limp from the wrists.

As the swinging continues with increasing amplitude you will find eventually that the arms will be above the head on the up-swing. Stop when the arms are in this position. Pause for a second or two then let the arms drop to the sides of the body under their own weight as in Exercise 1.

EXERCISE 3: THE SIDE-TO-SIDE SWING

Stand with feet slightly apart, as before, with arms hanging at the side. As in the previous exercise, the aim is to develop a relaxed swinging motion. This time the motion is a rotary one from side to side. Turn the body slightly to the right but keep the feet stationary. Allow the legs to twist so that the movement is not restricted to the upper part of the body. Swing the arms across the front of the body while turning. Then turn to the left swinging the arms to the left. With each swing bring the arms higher and allow the elbows to bend on the up swing so as to keep both hands at the same height.

Continue until the hands are reaching a position well above the head. Stop with the hands at the height of their swing.

Move the arms laterally to centralise them above the body. Pause for a second or two then let the arms drop to the sides of the body under their own weight as in the previous exercises.

10   *Breathing*

Good breathing is important for two reasons. First, as explained in Chapter 8, slow, deep breathing is one of the ways that we control the symptoms of nervousness. The second reason arises from the fact that breath is the power source for speech. We tend to think of our vocal organs as being the speech producers but these do no more than cause vibrations in the output of air from our lungs. Resonance then takes place in various parts of the head and chest. Just as an organ or a set of bagpipes relies on a sufficient throughput of air, we need our lungs to be able to provide us with ample air for all circumstances. Taking the analogy of a wind instrument further, our vocal chords are like the reed of a clarinet or a bassoon. The reed is essential in producing the sound, but is useless without the air to power it and is very limited in capability without the resonance provided by the tubular cavity through which the modulated air passes.

People who sing know the importance of good breathing. In singing, the required sounds are defined closely in terms of the words, the timing, the pitch and the relative loudness. Breathing must therefore be strictly regulated in order to achieve the desired result. In speaking, the requirement is not defined precisely nor is it unique. There are many right ways of presenting the information and it would be wrong to rehearse a presentation following a rigid pattern in a way that a singer must do. This means that the need for good breathing is not so readily recognised in speaking as it is in singing. Nevertheless it is equally important. There is a lot that public speakers can learn from singers. Notice how a singer readily fills a large hall with his or her voice, without the use of a microphone and perhaps in competition with accompanying instruments. Notice also that singing standards are much higher than public speaking standards. It would be an interesting experiment to hold a business meeting at which all speakers would be required to sing their contributions! The effectiveness of communication would I suspect be improved.

Our aim is to develop a natural tendency to breathe slowly and deeply. We tend to breathe using only the upper parts of the lungs. This gives shallow breathing which is adequate with regard to our need for oxygen but inadequate for effective public speaking. Although the upper ribs are fixed to the breastbone and immovable, the lower ribs on either side are attached by cartilage and can be moved outwards to give greater lung capacity. Between the ribs where they fall away to either side is a muscle called the diaphragm. It is the diaphragm that controls this rib movement. Like any other muscle if the diaphragm has not been used very much it will be weak and it may even be difficult initially to persuade the brain to bring it into consciously controlled activity. This is why some exercising is recommended.

For a number of reasons we tend to breathe in through our noses. We may have been told that it is healthier, or we may, I suspect, be conditioned to some extent by being told as children to keep our mouths closed when not in use for speaking or eating. Whatever the reason it gives us a disadvantage when we come to speak in public. We need to draw a large volume of air into the lungs and we do not want any unnecessary restriction to the flow. It is better therefore to allow the mouth and nose to give access to the incoming air.

EXERCISE 1: FINDING THE DIAPHRAGM[10]

Stand with feet slightly apart. Use one hand to locate the space between the two sets of lower ribs. Place the palm of the hand over this area. Breathe in slowly and deeply through the mouth and nose and you should feel the diaphragm move outwards. Do not raise the shoulders as you breathe in. Breathe out slowly and the diaphragm should return.

Now pretend that you have caught the scent of something pleasant. Take a series of sniffs as if to identify the smell, filling the lungs further with each sniff. You should feel a movement of the diaphragm with each sniff. Breathe out slowly.

Breathe in slowly and deeply again and then pant like a dog. With each pant the diaphragm should be felt to move. Breathe out slowly.

If at first no movement of the diaphragm is felt, you should not despair. It is likely to take some time practising the exercise until a ready ability to control the diaphragm is developed.

EXERCISE 2: DIAPHRAGMATIC BREATHING[10]

Stand with the feet slightly apart. Place the hands on the hips, fingers forward and thumbs to the rear. Keep the shoulders back without standing stiffly. Slide the hands forward slightly and upwards until the fingers are over the lower movable ribs. Now breathe in slowly through the mouth and nose without raising the shoulders. Feel the ribs moving outwards against the hands. When the lungs are comfortably filled stop and hold the breath for two or three seconds. Then slowly exhale while counting aloud until the lungs feel comfortably empty. Repeat the exercise for two or three minutes.

Notice that the lungs should be filled and emptied to a comfortable state. No straining or bursting of blood vessels should be involved! The counting can be used as a measure of progress but the exercise should on no account be treated as a challenge to see how far you can count using one breath. With practice the comfort level will of course change and represent increased depth of breathing.

EXERCISE 3: PACED BREATHING[11]

Stand with the feet slightly apart and with the arms hanging in front of the body, fingertips touching. Lean the head forward slightly. Breathe in and raise the arms sideways away from the body. Continue the intake of breath as the arms go above the head. The fingertips should meet above the head as the breathing in is completed. The head should be slowly tilting backwards as the arms are raised. Keep the arms relaxed and slightly bent at the elbows throughout, and keep the hands relaxed at the wrists. Breathe out as the arms are separated and lowered to their original positions. Tilt the head forward slowly as the arms descend.

The important point to note is that the breathing should occupy the full time that the arms are in motion. Thus, the speed at which the arms are moved regulates the speed of breathing. At first be content with fairly rapid movements, concentrating on making the breathing correspond exactly with the arm movements. Then gradually decrease the speed until the arm movements can be used as an automatic control of breathing rate.

11 Material

A method of preparing material for presentation as a speech or talk has been given in Chapter 4. In this chapter we shall discuss how it is possible to improve speeches by modifying the material. Clearly, the basic message cannot be altered: this is determined by the aims of the speech and the information to be used to achieve these aims. It should be remembered, however, that audience interest is vital. There is no point in presenting an accurate message if no one is absorbing it. It is therefore justifiable to use techniques which add to audience interest, provided of course that these do not detract from the intended message.

Good speeches, as opposed to adequate speeches, need considerable time for preparation, but time can in fact be saved by starting earlier. If you start collecting your ideas and information at the earliest opportunity you will find that you save time. This is because your mind will reflect in odd spare moments on the material you have collected. I suspect that some reflection may even take place subconsciously. The advice of 'sleeping on an idea' before finalising it, is perhaps an example of what I have in mind. An occasional glance through your rough jottings over a period of preferably several weeks will be worth much more than several hours work the night before the event.

As you collect your ideas always be on the lookout for general information that can be used to increase the interest of your audience. Obviously you need to have a link with the technical content of your speech. Historical aspects can be useful in this respect. Does your subject have any links with the Romans, or the ancient Egyptians perhaps? Are there any early inventions or processes that relate to your topic? Watch out for incidents or experiences in everyday life that provide examples of or analogies with the points you wish to make. Proverbs or quotations can be used, but be very selective. Apt quotations can often be found in unexpected sources but avoid anything, however apt, that has been overworked. 'Lies, damned lies and statistics' has become a cliché at technical conferences, and audiences are getting tired of hearing of developments or projects that 'growed—just like Topsy.'

Nursery rhymes and children's stories can provide links with technical subjects. Humpty Dumpty illustrates the effects of impact. The wasteful progression round the table of the Mad Hatter's Tea Party can relate to pollution. Agriculture is dealt with by Little Boy Blue and Mary, Mary, Quite Contrary. Lessons in building construction are given in the story of the three little pigs and in 'London Bridge is Falling Down'. Astronomy and space exploration are alluded to in 'The Man in the Moon is a Friend of Mine', 'Twinkle,

Twinkle, Little Star', and 'The Cow Jumped Over the Moon'. The seven dwarfs were involved in mining while Snow White and her problem with the apple prompts thoughts of forensic science. Numerous other examples can readily be found and the search can extend to well-known legends, folklore and mythology, and to songs and poems.

The search for suitable items is made easier if you start collecting now, in advance of specific requirements. In other words, use a notebook to record anything you come across that could be potentially useful. Appendix 5 gives lists of unusual items, classified by field of science or engineering, which can be used to start your collection.

It is useful to have access to books of quotations and general knowledge.[12-25] In the last category, for example, is the Guinness Book of Records.[18] This can be used to provide unusual facts on a variety of subjects. Many surprising statistics can also be extracted and, although audiences can find statistics boring at times, an unexpected statistic can heighten interest. It is quite likely that you have in your office one of those tear-off calendars that many firms provide. At the bottom of each leaf is a little piece of wisdom. The quality of these epithets is variable but it is worth a daily glance to see if there is anything worth noting for possible use.

Having mentioned quotations I should at this stage give some guidance on the way they should be introduced in your speech if you choose to use them. If the quotation is a well-known one, do not refer to the source. If you were to say, for example, 'As William Shakespeare said "All the world's a stage, . . ."' it would sound as though you were talking down to your audience. In a case such as this leave out the mention of Shakespeare. On the other hand if the quotation is not well-known you will probably have to give the source if only to let the audience know that it is a quotation and not something you have just made up. However, there is again a danger of appearing to talk down to the audience. You cannot say for example 'As John Sheffield, First Duke of Buckingham and Normandy said, ". . .".' It would be much better to say something along the following lines. 'Ever heard of John Sheffield? Probably not. I hadn't until recently. Who he was or what he did, I have no real idea, but he did say something of interest . . . .' Alternatively, you might make it more anonymous and briefer in the following way. 'Some three hundred years ago an English Duke said ". . .".'

The difficult decision arises when you are not sure whether your quotation falls into the well-known or the unknown category. It depends, of course, on the particular audience you are speaking to. Above all, avoid appearing to be clever. Indicate or imply in some way that some members of the audience may know the quotation better than you do: 'Some of you will be familiar with the words of . . .,' 'Those words may be familiar to some of you.'

From collecting material we have strayed into the writing of your draft speech. Let us now look in detail at the writing stage of your preparation. Because of your training you will tend to adopt a standard structure and this will be the structure you adopt in preparing written documents. Background information will be followed by details of what has been done. The findings or results will be followed by discussion, and conclusions and recommendations will conclude your draft. Before taking this for granted and setting pen to paper, consider experimenting with changes in the order of the material.

Adoption of a non-standard order of material in speeches can readily be justified. A speech, unlike a document, is not used for reference. It has a passing existence as it is presented and then exists only in the minds of those who have listened and remembered. A listener, however does not remember the points of a speech in the order in which they were presented. His memory is more like a file card system than a book. He can draw out remembered parts and put them together in the way that suits his own purpose. The speaker's job is to ensure that the parts are remembered, not that they are presented in a standard order. One way of ensuring that they are remembered is by maintaining the listener's interest, and one way of maintaining the listener's interest is by adopting a non-standard order of presentation.

An analogy is provided by a comparison between a history text book and a novel. The text book follows a chronological order so that students can readily find the particular information that they need. The chronological order is logical and therefore assumed to assist in the study of the subject. The novel on the other hand is not used for reference: it is intended to be read once only. The author adopts a technique of flashback and scene changing, purposely avoiding a logical development of the story. In this way he heightens the interest of the reader and encourages him to continue reading. I doubt whether the reduced degree of logic compared with that of a chronological account reduces in any way the reader's understanding of the relationships between different events.

You should therefore plan an order of presentation that maintains interest. Interest will be high at the beginning provided that your opening has caught the attention effectively. Make use of this by inserting important points early. You might even start with your conclusions or recommendations. Finer detail of your argument can come later when perhaps only those particularly in need of this information will be concentrating fully. Interest will be revived if you periodically summarise what you have said so far. It will also be revived if you use flashback or a change of scene. Signal the conclusion of your speech by a suitable phrase, 'in conclusion . . .' or 'finally . . .,' for example, and then make use of the revival of interest that this will bring. Conclude with a message that you most strongly wish the audience to remember. If your listeners remember anything at all about your speech it is most likely to be the opening and the conclusion.

If your speech is accompanied by a written document or paper that each member of the audience has been given, you have even greater scope for improving your speech. The most unsuitable procedure you could possibly adopt in such a situation would be to read the document. More likely you would prepare a précis of it. But best of all is to write a speech quite independently of the written document. It is fair for you to assume that details of your procedures, calculations, arguments or whatever can be extracted from the document by those who need the information. Your speech need not be burdened by such details. The role of your speech is to promote interest in the document by speaking about its contents, but not necessarily to relate its contents. The distinction is important. The speech must of course also create interest in itself otherwise it cannot promote interest in the document.

The distinction I am aiming to draw can perhaps be illustrated by comparing a précis of a book with a trailer of a film. The précis represents the conventional method of preparing the speech. The film trailer represents, albeit to an exaggerated degree, the approach I am advocating. It reveals the subject matter and the mood and style of the film. It selects certain details without attempting to précis the story. It is assembled in an interesting way and it raises interest in the film that it is encouraging the viewer to watch.

12   Presentation

The presentation of your material consists of how you look and how you sound. How you look depends on your stance and ways of improving this were covered in Chapter 9. In this chapter the use of your stance will be integrated with your vocal presentation.

Diction is the word used to describe the total vocal manner in which you present your message. There are many books dealing with diction much more thoroughly than we shall do here.[10,11,26,27] Diction is important to people other than public speakers, from actors and newsreaders to telephonists and British Rail station announcers. Many of the terms used by various authors are not uniquely defined and this can lead to confusion. It will be useful therefore to start with a number of definitions that will be used in the ensuing discussion.

The chart in figure 9 shows the relationships between the main constituent parts of diction. Pronunciation and articulation are concerned with the value the sound has in terms of language, whereas modulation refers to the different ways in which the same sound or quantum of language can be expressed vocally. The mechanical voices that we associate with robots, Daleks and suchlike devices have articulation but no modulation.

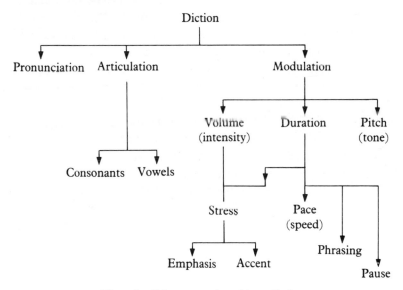

Figure 9 Diagrammatic guide to diction.

Pronunciation is generally fixed by convention so we have little choice. However, pronunciation of a word is not defined absolutely but only relative to other words. Thus, several people will agree a pronunciation of a particular word but each will actually say something different. Articulation is concerned with the actual sound that we give to each vowel and consonant making up the word.

The same word or syllable can be spoken in different ways

by the same speaker and this produces modulation. The volume or intensity can be varied and allows stress to be placed on various words or parts of words. Stress on a word or words is called emphasis while stress on part of a word is called accent. The duration of the sound can be varied and this plays a part with volume in the production of stress. The duration of the sounds is related to the pace or speed at which the material is presented. Pauses, the absence of sound, are used in pace control and also for effect. Phrasing is the use of pace variations and pauses to group words in terms of ideas. Finally, pitch or tone is used in the same sense as the musical term 'pitch' (but not the musical term 'tone'), that is to say it refers to whether the sound is of high or low frequency.

The best way of improving your presentation skill is by reading aloud with visual communication. A general procedure to adopt in practice will be given first. Then the various aspects of diction will be discussed and some variants in procedure to cope with particular aspects will be described.

Reading aloud is an ideal technique because not only is it beneficial but it absorbs little time since we read anyway. The visual communication is important because we must always look at the audience when we speak. Start by choosing a short passage of, say, a few hundred words. It does not really matter what the material is at this stage though more will be said about choice of material later. Choose something that you are going to read anyway. Go through the passage using a pencil to mark off groups of up to four or five words which form convenient idea phrases. Split the whole passage into such phrases. A double stroke with the pencil can be used to indicate the end of a sentence. (An example of such splitting up of a passage is given in Chapter 6.)

Now take up your stance in front of a mirror. Hold the book in one hand and locate the index finger of the other hand at the beginning of the passage. Hold the book high but check in the mirror that you are not obscuring your face. Take two slow deep breaths while looking at the first marked phrase, look up at the mirror and speak out the first phrase. Glance back at the book to pick up the second phrase, look up and speak. Continue in this way using the index finger to keep the place so that the eyes do not have to search for the next phrase.

At first the effect will be jerky. With practice you will develop a smooth presentation. It will seem slow but this should not worry you. In fact you should not proceed at a pace greater than 100 words per minute. You should check this by timing yourself over a known number of words.

Eventually you will be able to dispense with the pencil marks, and will in fact be able to read unseen material effectively while maintaining the required visual communication. At this later stage you will find that you are in fact glancing at the text quite a long way ahead and picking up signals as to when the main verbs, for example, will appear, or whether you are entering a lengthy subordinate clause.

If you have a tape recorder you will get some benefit from playing back recordings of your readings, though this is not essential. The playback will allow you to note any particular deficiencies with regard to diction.

Considering diction in more detail we can start with a few words about pronunciation. It is obviously important to use correct pronunciation and any doubt should be removed by consulting a dictionary. With names of people and places it is more difficult but you should nevertheless check in advance. In an emergency decide on a pronunciation, present it confidently and keep to it. Avoid apologising, stuttering or trying several versions.

Regional or foreign accents affect pronunciation because the reference sounds against which pronunciation is defined are different. It should be remembered that public speaking does not have the same aims as elocution. The criterion for acceptability of a public speaker is that what he or she is saying should readily be understood by the audience. If therefore you have a regional or foreign accent do not waste your time trying to develop a standard English pronunciation. Provided that people readily understand what you say, devote your enthusiasm to the other forms of practice detailed in this chapter.

Obviously, regional dialect falls into a different category from regional accent and would not be understood by audiences in general.

Articulation is concerned with the way that you execute your pronunciation. Thus pronunciation is in a sense mental whereas articulation is practical. Poor articulation gives speech that lacks clarity. The listener may understand what is being said but will find it an effort to concentrate. He may even miss much of the detail because many words are misunderstood. Most problems arise because of insufficient movement of the jaw, lips and tongue. We are inhibited to some extent when it comes to opening our mouths in public, probably a result of instruction in childhood not to exhibit our tonsils to visitors.

Considerable practice is necessary to improve articulation. The best form of practice is again reading aloud. Charts of vowel and consonant sounds are given in many books,[26-28] and repeated reading of the sounds in sequence develops a flexibility of the jaw, lips and tongue and a clarity of sound. Such charts are used by students of drama and their value cannot be overrated. For the busy scientist or engineer, however, I would recommend the reading of tongue twisters. These are readily available[29] and have the advantage of highlighting problems and presenting a challenge. In reading a tongue twister, take it initially at a very slow pace. Give every syllable its proper sound, aiming for clarity not speed. Consider the following example:

'Are you copper bottoming 'em my man?' 'No, I'm aluminiuming 'em mum.'

The difficult part is '. . . aluminiuming 'em . . ..' Try splitting this up using pencil strokes, thus

'. . . al/u/min/i/um//ing/em . . .'

Notice the double stroke after 'aluminium': you will find this useful as you increase the pace. Mentally visualise a pause at the double stroke as you approach it, but having reached it continue without a pause.

Modulation, as explained, is concerned with the ways we can vary the sound while articulating the vowels and consonants in a consistent way. The first contribution to modulation to consider is volume. Volume depends on breath and the ways of achieving good breathing have been described in Chapter 10. By varying the volume we can produce stress. Whereas accent is determined by convention, emphasis is used by the speaker as required. Emphasis is important in a speech, not only for the obvious reason but to give life and sincerity and to hold the interest of the audience. Emphasis should be used selectively: a particular fault of some speakers is to use emphasis in every sentence. It then loses its impact and the speech sounds like a badly recited poem. Emphasis is an effective means of signalling the end of your speech and making sure that your final statement is remembered by your listeners.

The way to achieve correct pace is by reading aloud and timing each attempt. The average number of words per minute should not be greater than 100. A constant pace, however is not what is required. Pace variations are essential to express mood and to highlight particular phrases. A change of subject requires a slower pace to allow time for the audience to adjust mentally to new ideas. If you slow down for unfamiliar or difficult phrases and speed up for everyday expressions you will increase the understanding and interest of your listeners. 'At the end of the experiment we measured the changes in temperature and weight of the specimen.' In a sentence such as this the pace can be relatively fast initially. At 'measured', the pace should begin to slow down and reach a minimum for 'temperature and weight of the specimen'.

Much is written about the use of the pause. With skill the pause can be one of the most effective devices a speaker can use. Pauses should be used to give time for ideas to sink in. They can also be used with great effect to create a feeling of expectancy. You will begin to use pauses with effect only when you identify the relative difference between how long a pause seems to the speaker and how long the pause seems to the audience. The conversion factor probably varies from person to person but the pause always seems longer to the speaker than it does to the audience. It is fear of the pause that leads to the habit of inserting 'um' or 'er' between phrases when speaking in public. If you suffer from this habit, cure it by attempting to substitute equivalent periods of silence, not by attempting to move on immediately to the next word. It is not easy to devise means of developing a skilled use of the pause. A lot can be learned of course by observing speakers in action but the following device will be of some help. Prepare a passage for reading and write in the word 'pause' at each full stop and other places where you feel a pause might be effective. Read the passage aloud saying the word 'pause' to yourself each time you encounter it. A playback of a recording of your reading will then start to give you a feel for the use of the pause.

Variations in pitch are necessary to express shades of meaning and should also be used to convey mood. Such variations are used naturally for these purposes in everyday speech but there are other uses of pitch variation for the public speaker. Because the pace is slow and pauses are used, it is important not to complete the sense of what is being said too soon. A descending pitch denotes ending whereas a rising pitch denotes there is more to come. Because the pace is slow there is a tendency to use a descending pitch at the end of each phrase. This tendency must be overcome. The sense must be kept open by a level or rising pitch. It is by keeping the sense open that a pause can be used to great effect without being taken as an ending. Compare the following two sentences:

'A stitch in time saves nine' is a popular proverb.
A popular proverb is 'A stitch in time saves nine.'

Read the second one aloud and you will notice that you say 'nine' with a descending pitch. This signals the end of the sentence. Now read the first one and you will notice that 'nine' carries a rising pitch because there is more to follow. Read the first one again with a pause after 'nine'. Repeat several times, each time increasing the length of the pause. As the pause gets longer resist the temptation to give 'nine' a descending pitch. This is the technique to use when carrying out your reading exercises. Approach the end of your phrases with the feeling that there is more to come, if in fact there is. It will then not matter just when you choose to continue. As far as the audience is concerned the sense is open and attention is maintained.

Many exercises have been devised to improve the various aspects of modulation. Most, however, are more suitable for a classroom situation than for self-tuition. In my experience the best and quickest way to achieve a degree of improvement is by adopting a state of intense awareness of the meaning of what is being said. This can be illustrated by means of the following sentences.

Mary said 'Sally is the best typist in the department.'
'Mary', said Sally, 'is the best typist in the department.'

In the two sentences the words are exactly the same but the meanings are quite different. In the written form, punctuation is used to give the different meanings. In speech we have to use modulation to give the different meanings. Notice that the articulation is the same for the two sentences: a robot or a Dalek would be unable to distinguish the two! Try reading the two sentences. Before starting each one, make yourself intensely aware of its meaning. Who is the best typist? Who is speaking? Now read as if your life depended on you making it absolutely clear to your listeners what the facts of the matter are. You will find your voice resorting to changes in volume, pace and pitch as required, provided you maintain an intense awareness of the meaning as you read.

A further useful example is given by the following sentence.

John, whereas Jim had had 'had had', had had 'had.' 'Had had' had had the teacher's approval.

Even with the correct punctuation, the meaning may not be

immediately evident. The sentence has been devised on the basis that John and Jim were involved in writing a passage for their teacher. John used 'had'. Jim used 'had had'. The teacher considered that 'had had' was correct.

Make sure that you are clear as to the meaning of the sentence and then try reading it aloud. The only way that you will convey the meaning is by being intensely aware of the meaning as you read. You will find that the intense awareness will produce the necessary modulation. The eleven consecutive 'had's each take on their essential individuality by variations in pitch, pace and emphasis.

Everything you read should be read with a feeling of intense awareness of the meaning. When you adopt this feeling you will notice immediate improvements in modulation. You can use the above sentences prior to starting your reading session in order to give yourself a reminder of the modulation that you can achieve.

An alternative way of getting your modulation working is by reading your first sentence with your mouth closed and the lips together. We are of course discussing practice sessions and not the real presentation! With your mouth closed your articulation will be so limited that the words will not be recognisable. But you will be able to use modulation. The challenge is to express as much meaning as possible by variations in volume, duration and pitch.

This chapter on improving your presentation will be concluded with a further mention of singing. The ability of singers in relation to breathing and volume of sound was discussed in Chapter 10. Singing offers a further example in relation to diction. In a song the pace and pitch are controlled by the music. This gives rise to a further useful means of practice. Before reading your selected passage, try singing it. Imagine that you are performing in an opera. Sing the passage making up the tune as you go along. Compose the tune to give suitable feeling to the words. At the end of your aria read the passage aloud, and you will notice some carryover of pace and pitch into your spoken words.

13 Notes

Good notes are essential for a good presentation. It is however, not easy to recognise good notes. The action of triggering the brain to produce a particular output of information, by means of a few words on a card, is no doubt complex and personal. The notes that would suit one person would probably not suit another. Furthermore, it is not possible to compose notes in a way that will guarantee that they are suitable even for the person composing them.

The only way to produce good notes is to subject them to a process of modification based on trial and error. This point was made in Chapter 4 where a procedure for preparing notes was given. In this chapter the means of becoming more skilful at making notes will be described so that the trial and error process can be shortened. In the two exercises to be described the use of notes is incorporated, so that improvements in both the preparation and use of notes can be achieved.

For the first exercise select a short article from a newspaper or magazine. Read it through absorbing the meaning. Read it through a second time underlining what you consider to be the key words or phrases. Copy out the underlined parts in the form of a list on a small card.

Now present the article out loud using only the list of key words and phrases. At the end of your presentation look back over the list and analyse the effectiveness of the various entries. Some of the words and phrases you will have found satisfactory in prompting a suitable response. Some will have been superfluous and should be deleted. Many, however, will not have been satisfactory and will have caused some struggle and discomfort in your presentation. These should be examined in relation to the original article with a view to alteration. A common fault you will find is the excessive use of nouns as key words. Adjectives and verbs are often more important in prompting the correct recall. Notice that it is frequently the mood or attitude of the message that has to be recalled rather than the factual content. This means that often the most useful key words or phrases are, for example, 'however', 'on the other hand', 'nevertheless', 'fortunately', 'if only' or 'regardless'.

Having made appropriate changes to your list, try the presentation once again and make further changes as necessary. The notes you end up with will be suitable for your use but would probably be inadequate for anyone else attempting to make the same presentation.

For the second kind of exercise you can again use newspaper or magazine articles but it is somewhat easier if you select a typewritten document. As before, read through and underline key words and phrases. Somewhat preferable to underlining is the use of a pale-coloured broad-tipped felt pen to paint across the selected words. In this exercise the key words and phrases are not transferred to a card but are used direct from the original document.

Now present the material out loud 'looking' at the imagined audience and glancing down merely to spot the coloured or underlined pieces. Surprisingly, this is much more difficult than the previous exercise. The temptation is to study further words and this leads to mental blockage so far as composing your own words is concerned. You may even find that your presentation degenerates into a reading of the passage.

Try again. The aim is to look only at the marked words and from these prompts to compose your own words to give the required message. As before it will be necessary to change some of the key words or phrases before a satisfactory presentation can be given.

This second exercise is particularly useful in that it not only improves your skill in making and using notes, but it gives practice in speaking from a full text. The technique of speaking from a full text, not to be confused with reading a full text, was described in §6.2.

14 Impromptu Speaking

Although it is true that some people seem to have a natural gift for speaking well without preparation, do not be misled into thinking that you cannot improve your skill at impromptu speaking. As with other skills, improvement is achieved by practice and by suitable exercises. Exercises take time and for the first of the exercises to be described you will need to allocate half an hour or so of your valuable time. The good news is that two further exercises are then described and these will not occupy any of your useful time.

For the first exercise choose a subject at random. Sit down with pen and paper and assemble your thoughts on the subject by means of a brain pattern as described in Chapter 6. From this, list the points you wish to cover and prepare notes. Allow yourself a total of 15 minutes preparation time and then make a presentation in front of a mirror. Aim at three minutes presentation time. Take a break for two or three minutes and revise your notes where necessary. Then repeat your presentation.

The value of this exercise is that it not only provides practice in collecting ideas quickly and structuring a speech at short notice, but it starts to give a feel for the relationship between message content and delivery time. One of the biggest faults to be seen in speakers is their lack of appreciation of the passage of time once they have started to speak. Remember the rule that was given earlier of a speaking rate of not more than 100 words per minute. This means that you can make not more than about eight short statements per minute. In a three-minute presentation this gives a maximum of 24 statements. A single idea, argument or description could easily require three or four statements and we can begin to see how quickly the time will pass. The first time you try the exercise you will find that you have attempted to assemble far more into your communication than your time of three minutes allows. With practice you will begin to be aware of the time required to express your ideas.

The second exercise uses the 'commentary technique'. You can carry out the exercise during any odd minutes that crop up while you are alone. Driving along in the car, lying in the bath, waiting for the kettle to boil, jogging or walking through the countryside; these are some of the occasions when you can make valuable extra use of your time. Imagine you are a commentator giving an account of what you can see around you. Start without preparation but in a formal style:

'Ladies and Gentlemen. As I approach the crest of the hill along this narrow winding path, the tops of distant mountains come into view. After two hours walking I am looking forward to a brief rest. I can see ahead a slab of exposed rock which promises a suitable place to sit for a few minutes . . . .' Continue in this fashion maintaining a flow of words, initially without worrying too much about the content. As your talk becomes smoother concentrate on the structure of your statements. Before coming to a close plan your ending. '. . . But above all it is the sight of the river meandering along the floor of the valley—a sight which for me will always be Whitchurch.' Avoid ending abruptly and avoid petering out.

You can use the technique while you are carrying out tasks such as washing up, gardening or building models of the Houses of Parliament out of matchsticks. Your commentary will then be a description of what you are doing. Make an effort to avoid long gaps. Fill in with background information. Police drivers use a technique of 'commentary driving' in which they describe aloud their actions and their plans in relation to the road ahead, traffic, pedestrians, etc. The technique is designed to improve driving skill, but equally it will improve speaking skill.

The third exercise makes use of time that would otherwise be wasted. It may be called the 'just-in-case' technique. It is based on the concept that wherever you go you should imagine that you may be called upon to speak. If you go to the cinema or theatre you will sit waiting a few minutes for the performance to start. Instead of day-dreaming, imagine that you will be called upon to address the audience. Perhaps the projectionist has been taken ill and there will be a short delay. You have been asked by the manager to inform the audience. Start mentally composing a suitable address. It will not be needed but it will be good practice.

When travelling by bus or train you can use the same technique. Again the planned speech is hardly likely to be used, but there are occasions when you could be asked to say something. At dinners, dances and other social gatherings someone could be planning a surprise for you. Think of those people featured in *This is Your Life*, a well-known TV programme that brings unsuspecting people before the cameras. If you adopt as a matter of principle the thought that wherever you are you could be asked to address those present, then you will be prepared. The mentally composed speeches will not only provide practice but will also give you insurance against the need for such a speech arising. Clearly you cannot envisage the required content of such a speech with any accuracy, but you can get clear in your mind much of the background to the event. This would at least provide a suitable introduction and would create thinking time for the preparation of your message.

15   Observation and Analysis

It hardly needs pointing out to readers with a scientific background that there is value in observation. You can learn much about the skill of public speaking by observing the performance of others and of yourself. From observation arises the possibility of analysis, and it is this that distinguishes between casual observation and observation directed to the gaining of information that can be put to practical use.

Observation of one's own performance, being subjective by its nature, is not easy. It is therefore necessary to attempt to compensate by using a scheme of observation and analysis designed to be as objective as possible.

A good arrangement is to have your real-life performance tape recorded but this is not generally possible for various reasons. A video recording of the event would be ideal but this would be out of the question for most of us. You can of course use a tape recorder during your practice sessions but you must remember that your practice performance will not be quite the same as the real-life presentation. The choice is very much the same as the choice that faces us in our scientific investigations. Should we observe the real situation and suffer problems of making the observations, or should we set up a simulated situation which may be no more than approximate but which eases the problem of observation? In fact we usually choose to adopt both methods if possible.

Let us assume that you have produced a recording of your performance. The task now is to identify faults and good features. Appendix 3 consists of a checklist suitable for this purpose. The list, in attempting to be as comprehensive as possible, includes visual aspects. This may present some difficulty if you are working from a sound recording only, but it is possible to use the entries on the list to prompt useful recollections. This same checklist can of course be used to analyse recordings of your practice sessions.

Try to distinguish between two kinds of faults. The first kind is associated with the particular presentation while the second kind represents your personal characteristics. Stumbling over a word or forgetting where you are will be very noticeable during your analysis, but remember these faults are unlikely to recur. They are characteristic of the particular moments of time when they occurred so do not give them undue attention. More important in the long term, are the faults that are part of your natural way of doing things. These faults are obviously more difficult to recognise. The only advice that can be given is that you should attempt the analysis as objectively as possible. At least, with a scientific or technical training you have as good a chance as anyone of making an objective assessment of your own performance.

In relation to analysis of your own performance a word of

warning is necessary. It is of no use asking other people to comment on your performance. It can be a very sensitive subject and the comment you get will be influenced by the situation. Your superiors, though ready to criticise documents that you have written, will probably be reluctant to criticise your verbal presentations for fear of destroying your confidence. It is in any case an area in which they do not feel competent to offer constructive criticism. Superiors more frank in their approach may criticise, but their remarks are likely to be influenced more by the consequences of your efforts rather than by the skill displayed. Juniors are likely to be complimentary while colleagues are likely to give you answers that they think you would like to hear. I am afraid that this is a situation in which 'even your best friend won't tell you.'

A better way to judge your performance is by observations during your presentation and afterwards. Did you catch the attention of the audience when you started to speak? Did you sense a hush descend on the room or was there a mumbling and shuffling that took some time to die away? Did people look interested? Were any chatting to their neighbours, rearranging papers or even dozing off to sleep perhaps? During pauses, was everyone silent and still with eyes focused on you? Did you see anyone looking at his watch? (Not shaking it to see if it had stopped, I trust!) When question time arrived were people slow to raise points or develop a discussion? Did the questions reveal a misunderstanding of what you had been trying to say? How much time did you take over your presentation and could you have been more effective in a shorter time?

It is possible to test your effectiveness in some instances by judging the results of your presentation. At meetings that are minuted examine the draft minutes and note whether the secretary got the point of what you were saying. Consider how well you achieved what you set out to achieve. With hindsight, to what extent would you have changed your presentation to have made it more effective?

It is not only valuable to analyse your own performance but also to analyse the performance of others. You will have no difficulty in finding examples of poor performance. Any technical conference will provide examples of most faults. Committee meetings are useful in this respect but don't allow yourself to be distracted from the business of the meeting by analysing the speakers. Notice that it is not sufficient to judge a speaker to be poor, you must establish why. The checklist in Appendix 3 can be used to aid the analysis. Analysis should not be restricted to examples of poor performance. Watch out for speakers who are particularly effective, not only real-life examples but also speakers on television. Again, attempt to analyse why the presentation achieves a high standard. Note particular techniques that a certain speaker may use.

It is not necessary to embark on this analysis of others in a formal way. You will find that you begin to analyse automatically once you have become aware of the factors that contribute to a good presentation.

16   Classes and Examinations

It was mentioned in the previous chapter that there is no point in asking friends to comment on your performance as a public speaker. There is, however, one situation in which it is possible to get honest constructive criticism of your performance. This is the classroom situation.

By far the best way of improving your skill is by enrolling in a public speaking class. The benefit you obtain in this way comes mainly from the practice provided by the exercises set by the tutor, and by the comment and guidance that he or she is able to give you. In the course of preparing and delivering speeches in front of the class you will be able to experiment and observe the result. Such experiment is otherwise impossible. In the real-life situation every performance counts and experimentation could be risky. The control of correct pace, and the feel for the duration of effective pauses are examples of skills difficult to acquire without conscious experimental variation.

If you feel that lack of confidence is one of your difficulties you would find particular benefit from class tuition. Every student in the class is suffering from nervousness and the building up of confidence proceeds rapidly because of the mutual appreciation of each one's difficulties. Observation of the performance and improvement of others in the class is an important means of improving your own performance. You can measure the effectiveness of each other's experiments in technique.

Basic theory is of course taught, but this can readily be obtained from books. The value of classes is not in the formal lecturing but in the practical work, which usually takes up three-quarters or more of the class time.

Class homework usually consists of preparing short speeches. Free choice of subject is generally allowed so that it is possible to align the work to personal requirements. Thus, the scientist or engineer can gain valuable practice in talking about his work to an audience representing a wide range of knowledge and interests.

Evening classes in public speaking are available in most localities and your local library will usually have details. At a cost of a few pounds per term a series of classes represents good value for money. Evening classes are attended by a wide range of people. Some businessmen enrol to improve their skill in business meetings or in dealing with members of the public. Housewives who find themselves elected to the local Women's Institute committee are well represented. Some people come along because they have special occasions to face such as weddings or presentations. Some just come along because they are night-school addicts seeking a new subject of interest. Young and old are to be found in each class and for some reason that I do not know of there always seems to be at least one police officer. The variety to be found among the students gives an added interest and a degree of entertainment. It is surprising how fascinating it can be to listen to people from different walks of life speaking about their work and their interests. Classes in fact are good fun.

Some firms of course arrange their own public speaking classes, usually as part of courses dealing with communication skills generally. My own experience of such classes is that they are useful but usually too brief to have great effect. One or even several lectures on public speaking can do no more than identify the problems. Curing the problems requires each student to have short practice sessions spread over a number of weeks. Greatly contracted courses allow no time for each degree of improvement in skill to be assimilated and related to the real world. Firms that run their own classes would do well to consider, as an alternative, sending employees to evening classes and paying their fees. This would be much more effective and would probably be very much cheaper so far as the company was concerned. If your particular company does not run its own courses, why not enquire from your personnel officer as to whether you could attend evening classes at the firm's expense?

It is possible to take examinations in public speaking. Evening classes prepare students for examination but students are by no means obliged to take them and probably less than half do. Appendix 4 gives details of examinations in public speaking that are set by colleges in the United Kingdom. Many of these examinations can be taken in other countries and where this is so a suitable entry is included in the appendix. The examinations are essentially practical though the more advanced qualifications require additionally an adequate performance in a written examination.

The practical examination consists of several parts which vary according to the grade of examination and the examining board. In Appendix 4 the parts are identified by code letters which relate to the types of exercises the student is required to carry out. The types of exercises are described below.

P—prepared talk or speech

The student prepares a talk or speech in advance of the examination. The subject may be of the student's choice or, in the higher grades, will be selected from a list published in advance by the examining board. The time for the speech is specified and ranges from two minutes to ten minutes, depending on the examining board and generally increasing with the examination grade. Notes may be used by the student in the presentation. Visual aids are generally allowed and may be specifically required for some examinations.

I—impromptu talk or speech

The student prepares a talk or speech on a subject given by the examiner. The usual arrangement is for a short list of subjects to be passed to the student 10 or 15 minutes before the start of the examination. Having made a choice from the list, the student prepares suitable notes. The maximum time for the presentation is specified and ranges from three to five minutes.

R—reading

The student reads a short passage of prose or poetry. Usually the passage is selected by the examiner at the time of the examination and no preparation time is allowed. In some of the lower grades the student is allowed to prepare the passage in advance.

M—memorised presentation

The student presents a short piece of prose or poetry from memory. As can be seen from Appendix 4 this type of exercise is not commonly encountered.

O—oral précis

The examiner reads a short passage to the student who is then required to give an oral précis of the content. Again, this type of exercise is fairly uncommon.

D—discussion

This heading is used to group together a range of possible activities having the common feature of conversation between examiner and student. In some examinations the conversation is on general topics but more commonly it consists of questions and discussion relating to the student's speeches. In the higher grades the student is expected to answer questions on speaking techniques and theory.

There are several advantages in taking examinations, some not so obvious. The achievement of a recognised standard could obviously be of benefit in one's career development. Even if employers do not give formal acknowledgment to public speaking certificates, their attainment is a factor that will be noted in any consideration of employment or promotion. Probably of greater importance is the effect it has on one's confidence. To have gained an examination award gives a feeling of satisfaction which undoubtedly improves the speaker's performance by removing nervousness based on uncertainty of doing the right thing. In some examinations the examiner issues a report of performance, pass or fail. This gives a skilled and unbiased assessment, noting weaknesses and good features. The report in relation to the examination fee is value for money, even if the pass mark has unfortunately not been reached.

Part III

Particular Events

17 Speaking at Technical Conferences

A technical conference is probably the most formal of the occasions at which the average scientist or engineer will be required to speak. Preparation and presentation make greater demands than at less formal events and for this reason Part I of this book has been written very much with technical conferences in mind. It remains therefore in this chapter merely to draw attention to a number of additional points.

The audience may amount to several hundred people and the hall may correspondingly be very large. This means that a microphone will be used and visual aids will be limited to projectors of the slide or overhead type. The need to familiarise yourself fully with the microphone and the projector arrangements has already been pointed out. This is particularly important in the large conference situation.

You will have taken the opportunity of trying out the aids before the session in which you are to make your presentation. It is common practice however to have several speakers in one session follow each other without a break. Bear in mind that you cannot expect to find everything as it should be if you are following one or two other speakers. The microphone may be left switched on or switched off. As you look at the switch, will you recall whether up is on and down off or vice versa? Remember, you must not blow down or tap the microphone.

The obvious lesson is to take this into account when you have your try-out of the equipment. In addition, you should watch the previous speakers carefully. Notice where the remote control for the slide projector has been placed. Did the previous speaker replace it on the lectern or did he put it on the edge of the chairman's table as he walked over to the screen to point something out? What has he done with the pointer? Is it back in the corner where it normally stands or did he march from the stage with it like a drum majorette? Mentally check everything that determines the situation you will move into after your introduction. Are the curtains drawn? Is the focus adjusted satisfactorily on the projector? Will the microphone height need adjusting or is the previous speaker about as tall as you are?

As soon as you take up your position after your introduction you should locate and adjust everything so that it is how you want it to be. Do this before you begin to speak. This will ensure that your presentation is not spoiled by unnecessary interruptions.

Observation of all previous speakers at the conference will save you from possible pitfalls. You will be able to note the relative positioning of the projector, the screen and the speaker's lectern. If there is a tendency for speakers to block the screen from view you will take note and decide what to do about it. What other problems are the speakers having? How easily can they see the screen? Are their signals to the projectionist misunderstood? Is the pointer long enough? Is the microphone highly directional? Are there cables to be tripped over? Is the microphone height easily adjusted or is it very stiff? Do the floorboards creak as speakers walk to the screen? As you observe make a note of everything that distracts by looking or sounding not as it should be. Decide whether your presentation may suffer from the same problems and if so decide what you are going to do about each of them.

An important feature of conferences is the allocation of time to questions and discussions. Dealing with questions is an exercise in impromptu speaking and as such has been dealt with in Chapter 6. However, you are in a position at a conference to anticipate to some extent the questions you may be asked. Make a list of likely questions in advance and prepare replies. It gives a good impression if you are able to give full answers to specific questions so take along your detailed results and observations if possible. It may be possible to have with you additional slides or transparencies that can be produced in answer to questions.

If you are asked a question that you cannot answer, say so quite simply and avoid irrelevant comments. Ask the questioner to give you his address after the meeting and promise to write to him with the information or comment he requires. Do not assume that every point that is raised in discussion requires a comment from you. It is often appropriate merely to smile or nod an acknowledgement to the contributor and await the next question.

At large conferences, roving microphones will be available for the use of questioners. At smaller conferences this may not be the case. When microphones are not available it always seems to happen that it is the people sitting at the front who ask the questions. They face the speaker, instinctively, with the result that those to the rear of the room cannot hear the question. The chairman should repeat the question over the microphone so that everyone is aware of the point being made. If he does not, then take advantage of the situation and repeat the question yourself. This gives you valuable seconds of thinking time in which to plan your reply.

A few words of advice to contributors during discussion periods are appropriate at this point. Compiling a question or comment is an application of the techniques of Chapter 6 for speaking at short notice. Some comments of course can be prepared well in advance. It is not uncommon at technical conferences to allow contributions which are related to the theme of the session but do not specifically refer to any of the presented papers. Such contributions should be kept very short. Some contributors abuse the system by giving prepared presentations, often with slides or transparencies, which are almost as long as the scheduled presentations and which absorb valuable discussion time. If you wish to make a short presentation ensure in advance that it will be welcomed. It does happen that a chairman, having asked for a question and been subjected to a lengthy talk, will show his annoyance by a pointed remark to the contributor. He may ask for the question to be clarified!

Many technical conferences are international in nature. If you are making a presentation in a foreign country the question may arise as to whether your paper should be given in English or in a language less familiar to you. I cannot advise from personal experience but I can draw attention to a useful debate on the subject. In an article in *Physics Bulletin*[30] it was suggested that more of us should be prepared to present papers in foreign languages. The proposition was supported by an account of the author's own experience. He explained that, whereas his presentation had been well received, it had been said that little more than 20% of the technical content of papers was coming across to the audience when the simultaneous translation facility was used by other English speakers.

A checklist for those planning a foreign language presentation was included and a summarised version is given below.

(1) The attempt is probably justified only for a worthwhile contribution with first-rate contents.

(2) Provide yourself with ample preparation time.

(3) Well illustrate the lecture to shift attention from the spoken word.

(4) Make sure the English version is satisfactory before moving into the foreign language. Allow for speaking more slowly in the foreign language.

(5) Use a translator rather than attempting a preliminary translation yourself. Explain the technical expressions to him and get a second opinion on the whole translation.

(6) Check that the translator has not misunderstood the original nor misplaced intended emphasis.

(7) Run through with someone who can check pronunciation.

(8) If simultaneous translation facilities are to be provided, give copies of English and foreign texts to the interpreters.

(9) Many phrases will be memorised and this will allow the

eyes to be lifted, for example, to point to the screen. Mark the script where slides are to be changed. Learn the colloquial phrase for 'next slide please' if the projector is not to be operated by yourself.

(10) Make use of blank slides to avoid calling for the lights, etc.

(11) Instruct the projectionist that focusing etc is his responsibility. Provide him with a marked copy of the script and instructions to vary the room lights when required.

(12) Memorise the first sentences and the closing remarks.

(13) You will probably be able to understand the questions but reply slowly and clearly in English. Do not attempt to ad-lib in the foreign language.

The article produced some interesting comments in the Letters column of the journal. Boyce[31] spoke in defence of interpreters who have the difficult job of undertaking simultaneous translation. Michelson[31] and Hird[32] disagreed with the suggestion that speakers should generally attempt to deliver a presentation in a foreign language. Michelson suggested that the criterion should be whether the speaker's mastery of the language allowed him to understand and answer the questions. Hird pointed out that a higher level of fluency is demanded from a speaker than from a listener and offered three criteria to be met before the speaker should use a foreign language. These were essentially as follows:

(1) Does the speaker understand the foreign sentences, as he reads them, better than his listeners understand well-spoken English?

(2) Is the speaker confident that his accent and intonation in the foreign language are good?

(3) Does the speaker have a translator who is both bilingual and technically competent to understand the subject matter?

If the answer to any of these questions is no, then, it was suggested, the speaker should use English. Metaphors and colloquial expressions should be avoided and the delivery should be slow and clear.

It sometimes happens, again particularly at international conferences, that authors who are unable to attend invite other speakers to present their papers on their behalf. If you are invited to present someone else's paper take comfort from the fact that it is not as difficult as it sounds. In fact in some ways it is easier than presenting your own paper.

The first requirement is that you understand the paper, and it may be necessary to clarify points with the author. It is, however, not necessary that you understand any more than what the paper contains. As far as you are concerned the paper is the sum total of information on which your presentation will be based. In this sense the preparation of your material is easier than if it were your own paper. Write your own full text for your presentation rather than use a version prepared by the author. Then prepare notes in the usual way. It is always easier to précis someone else's written material than your own, because you do not cling to the carefully composed sentences which seem just right from the author's point of view. Rather, you view the paper as the audience would. What are the essential points? What needs to be explained? In a sense you act on behalf of the members of the audience extracting what they need.

You cannot of course deal with questions and it would be wrong to attempt to do so. What you can do is to undertake to communicate questions and comments back to the author for him to deal with subsequently.

18   Chairing Technical Conferences

The usual arrangement at technical conferences is to appoint a chairman for each session, a session generally being a period between two breaks. A day will typically be broken into four sessions, with morning coffee, lunch and afternoon tea providing the three breaks. Each session will include several papers, perhaps four or five, each presented by a different author.

The chairman's obvious role is to introduce the speakers to the audience and to control the proceedings during the discussion period. In order to do the job well, however, it is necessary for the chairman to appreciate the need for him to present the session as an entity. The audience must feel that progress is being made and not that the session is a verbal scrapbook of bits and pieces. The chairman needs to identify themes and provide continuity.

If you are faced with undertaking chairmanship of a session at a technical conference, you will generally find that the organisers have attempted to identify a common theme for the session. Do not assume that this has been done effectively. Remember that the organisers had to plan the conference before they had seen the final versions of the papers, perhaps before many of the papers had been written. Your first task therefore is to obtain copies of the papers, to examine them and establish common themes, progressions and contrasts. You are then in a position to see your session as a unified part of the conference and to prepare your introduction, continuity and closing remarks appropriately. Give some thought to the discussion period. Simply waiting to see what questions or comments turn up may be satisfactory, but it may not be. It is much better to identify areas where questions would serve a useful purpose in developing the theme of your session. Areas of agreement or disagreement between authors can be identified. Such preparation will allow you to guide the discussion session along productive paths.

Arrive early for your session so that you can carry out a number of tasks before the speakers arrive. Confirm with the organisers that there have been no last-minute changes to the programme. It does sometimes happen that a speaker is ill and a substitute is making the presentation, but no-one has remembered to tell the chairman. Familiarise yourself with the geography of the building. If the coffee break follows your session for example, you will need to inform the audience where coffee is being served. Examine the conference room noting the seating arrangements. Where will you be sitting? Where will the speakers stand to make their presentations? Where will they be sitting while waiting to speak? Try the microphones and projector controls. Find the light switches, pointer, chalk, etc. Discuss with the organisers the procedure to be adopted in the event of a failure of the electrical equipment. Keep in mind during these preliminary checks that if anything goes wrong during your session you will be the person expected to take the initiative. Consider all possibilities: a speaker could dry-up, or faint perhaps, or the fire alarm might be sounded.

Meet the speakers before the session starts and take them to the conference room while it is still empty. Make sure they understand where to sit and stand and how to operate the microphones and visual aids. Deal with any queries they may have, referring to the organisers as necessary. Give each speaker an outline of your prepared introduction and be particularly careful to ensure that you know the correct pronunciation of his or her name. Discuss with the speakers the themes you intend to highlight and draw attention to similarities and differences to be found in their papers. It would not be unusual to find that the speakers have had little time, or inclination perhaps, to study each other's papers.

It is important to emphasise the need to keep to time. Arrange a system whereby you will signal the speaker one minute, say, prior to the end of his allocated time. The signal should preferably be one that is not obtrusive but one which is noticed by the audience and not just by the speaker. In this way you ensure that not only does the speaker know he is due to finish but he knows that the audience knows! A small card passed across to the speaker is probably the best signal that you can generally use. A better arrangement, but not one that I have seen used, would be a coloured light visible to both speaker and audience. Some lecterns are equipped with a signal light visible only to the speaker. This is not satisfactory as many speakers ignore it and can always claim afterwards that they did not notice it. One good signalling arrangement is based on the chairman leaving his seat when the presentation starts and returning a pre-arranged number of minutes before the end. During the presentation the chairman sits with the audience in the front row. In this way he has a clear view of the speaker and the screen and avoids a crick in the neck from attempting to see the illustrations from his usual position. His paced return to the chairman's seat is unobtrusive but provides an effective signal to both speaker and audience.

Having considered the preliminaries we can turn to your presentations. Your opening remarks are important and may determine the success of the session. The need to catch the attention immediately in any presentation was stressed in Chapter 4. One of the most difficult situations in which to catch the attention immediately is in opening a session at a technical conference. People attending a conference are as interested in meeting and talking to each other as they are in listening to the speakers. You will have to interrupt many private conversations in order to start on time. If your session follows a short break, you will have the added difficulty that people will be filing into the hall minutes or seconds before the start and then there will be the inevitable latecomers. You need therefore to apply the techniques given in Chapters 4 and 11 in order to ensure an opening that brings the audience to an immediate state of interest and anticipation.

In your opening address you should welcome your speakers, identify the theme of the session and indicate its relation to other sessions of the conference. You may need to explain certain administrative arrangements such as meal times, transport, registration procedures, etc. Explain also how questions and discussion will be dealt with. There may be time allocated for questions after each presentation, or there may be simply a general discussion period at the end of the session. Sometimes both are scheduled. If a roving microphone is to be made available for contributors from the audience make this clear.

Request written copies of each question or contribution if these are required by the organisers for inclusion in the published proceedings. Ask each contributor to start by giving his name. This helps the speaker to follow up anything arising from the contribution.

As you introduce each speaker strike a balance between saying too little and saying too much. Say who the speaker is and indicate his area of expertise in relation to the subject to be covered. Link his paper to the theme of the session and to what has gone before. Resist the temptation to say too much and, whatever you do, do not anticipate what he is going to tell the audience. He is the presenter of his paper, not you, and the last thing he wants is for you to give the audience a summary of his paper.

While the speaker is making his presentation you will be in full view of the audience. Courtesy demands that you pay attention and do not distract the audience. The worst examples of behaviour are provided by the chairman who moves from his seat to whisper a message to someone sitting on the front row, or passes written messages to members of the organising team. I suspect he does these things in order to appear to be busy and in control of everything. Look at the speaker periodically and make notes which will assist you during the discussion period. Observe the audience and judge the degree of attention being given to the speaker. Watch the time and signal the speaker exactly as prearranged. You will be tempted not to signal particularly if the speaker sounds as though he is about to round off, but do not be misled. The speakers who make a habit of overrunning their time, develop a knack of seeming for prolonged periods to be near the end. This is a protective mechanism acquired to keep anxious chairmen from bringing down the curtain. Another source of temptation to abandon the signal arises if the speaker is the last one of the session and the discussion period is to follow. You may then feel that a few minutes extra will not really matter and, furthermore, if there are no questions you will look foolish having cut short the speaker. Again, resist the temptation. There will be questions, and your job is equally to ensure that the contributors from the audience are given their allotted time as it is to ensure that each speaker gets his or her allotted time.

When the speaker finishes, lead the audience in a round of applause. If you have to cut short the speaker there is no need to apologise. If you sympathise by telling the audience that the speaker had a difficult task in covering his subject in the allotted time you are making things worse. The speaker retires confirmed in his belief that keeping to time is impossible and not to be worried about. Thank the speaker and begin your introduction of the next speaker or the discussion period, as appropriate.

You will have prepared a number of questions and points for discussion from your original examination of the papers and from your identification of the theme of the session. Notes made during the presentations will have supplemented or perhaps corrected some of your items. You are therefore in a position to start the discussion if members of the audience are initially slow to contribute. You should retain control of the discussion throughout, not allowing questions or comments to be fired across the room with the speakers jumping in when they can. Take one question at a time and repeat the question for the benefit of the speaker, making sure that it is clear to which speaker or speakers the question is addressed. Comments need not be repeated unless there is no microphone available in the audience and those at the back of the room may not have heard. Feel free to guide the discussion particularly if it has begun to focus on some controversial but minor point arising in one particular paper. You can always suggest that questioners take up the matter with the speaker after the close of the session, and you can set the discussion off in a different direction using your prepared notes. Have constantly in mind the theme of the session and its role in the conference. Conferences are often judged, by those attending, on the basis of the quality of the discussion periods rather than on the quality of the papers presented. Time and attention devoted by the chairman to preparing for and guiding the discussion is thus well spent.

Allow yourself a minute at the end of the session to make a few concluding remarks. Thank the speakers and contributors, and summarise in relation to the theme of the session. Your summary must be balanced and you must not at this stage add any new point or opinion which speakers or audience have had no chance to reply to.

19   Addressing Small Groups

Most of a scientist's or engineer's involvement in public speaking is likely to be to small groups of people up to, say, twenty or thirty in number. The occasions can range from office presentations to talks to parties of visitors.

The main difference between the small-group situation and conferences is the degree of formality. Less formality is called for in speaking to a small group but always aim for a little more formality than you think may be necessary. It is easy to move towards less formality if you need to, but it is impossible to raise the level of formality once you have made a start. Some speakers make the mistake of attempting to create an entirely informal atmosphere. This is done, subconsciously, to lessen the nervousness associated with speaking. The fear of public speaking is associated with the formality of the situation. The result of trying to be informal can be disastrous. The address can degenerate into conversation, the audience will interrupt and talk among themselves, and the speaker can wander from his theme. Timing goes astray and the event is anything but memorable.

As a final comment on formality it is worth noting that formality has nothing to do with seriousness. A presentation can be light-hearted or humorous and still be presented in a formal manner. It is the speaker's stance and appearance that first signals the degree of formality. The use of a formal opening, 'Ladies and Gentlemen', the slow pace and good volume and clarity of speech confirm the formal nature of the occasion.

In planning your address you must consider the nature of the audience. What is it that has brought them together as a group to listen to you? What have they in common? What are they likely to know already about this subject? What kind of information are they seeking? Is it detail or generalities that they want? Should the information be precise or simplified? Your evaluation of the audience needs to be more accurate when the group is small than when it is large. A larger audience generally represents a wider range of interests and prior knowledge, so someone somewhere is paying attention. A small group spoken to at the wrong level will show very obvious signs of distress and the invitation to question the speaker may be met with silence.

Visitors can range from students from the local school to a collection of university professors, or from junior trainees in the company to very senior people. It is wrong to imagine that your standard description of what your department does will suit each of these audiences. Each group of visitors requires its own presentation and your job is to prepare a presentation that matches the need. Find out in particular what the group has already been told by way of background to what you have to say. If you set the scene for your visitors by repeating what someone else has already said it will be irritating and a waste of time. It may also be embarrassing if your version differs from the earlier one.

Visitors usually arrive late if they are on a conducted tour. This is because previous speakers have not kept to time. You know by now the importance of keeping to time and your presentation will have been planned to meet the scheduled time allowance. This will give a long-term benefit in that you will become recognised as a speaker who keeps to time. You will find yourself selected more frequently to take part in the important events that arise. A speaker who keeps to time is appreciated not only by his audience but by those who have the task of organising events.

Prepare notes as explained in Chapter 4. You may think that you do not need notes in a situation such as this. After all, you know your subject thoroughly, you have dealt with similar events on many previous occasions and you would prefer not to suggest to your audience that you cannot speak without notes. It is exactly because you know so much about your subject that you do need notes. Without them, you will have too much to say. You will extend your presentation into various additional areas as your thoughts connect with other thoughts. You will be carried away with a fluent ramble through the wealth of your knowledge. You will lose all sense of time and have to be stopped in full flight. Use notes whenever possible, that is to say, whenever there is sufficient advance notice to allow you to prepare them. Never be ashamed of notes. Notes indicate to the listeners that you have gone to some trouble on their behalf and that what you are going to say has a beginning and, thankfully, an end.

If your visitors do not have a technical background, be careful to avoid using expressions that they will not understand. It is easy to identify the scientific terms particular to your field of work and to remember to explain to your audience what they mean. It is, however, much more difficult to appreciate that many of the common technical expressions can mean something quite different to the layman. 'Precipitation' means a fall of rain or snow, or a plunge from a cliff top, to most people. A 'plot' could be a sinister scheme or a corner of a garden. Words such as 'mass', 'element', 'base', 'mean', 'tension', 'stress' or 'strain' can cause confusion depending on the level of technical knowledge of your listeners. In addition, try to avoid using the expressions that have become clichés in discussions between scientists. Phrases such as 'this is not so', 'are not significantly different', 'less than or equal to', or 'can be readily shown' will be understood, but their use will create a style of speaking unfamiliar to the layman.

Visitors whose first language is not English require a similar sympathetic approach. Keep statements short and simple and make your delivery at a slow pace with ample pauses. If the services of an interpreter are being used you will need to stop periodically and await the translation. It is not easy to judge how frequently to stop. It depends very much on the ability and technical knowledge of the interpreter. The best technique to adopt is to watch the interpreter closely. Pause at the end of each statement to allow the interpreter to take over. If he or she does not show signs of being ready to speak, then continue with your next statement. Sometimes the interpreter will attempt a simultaneous translation, in which case it is particularly important to proceed slowly and stop periodically to allow the interpreter to catch up. Remember that the audience is listening to the interpreter and not to you. You will find that you will begin to speak louder when you hear the competing voice. Resist this instinctive reaction and maintain a volume no more than sufficient for the interpreter. Concentrate carefully on what you are saying or you will be distracted by the foreign rendering of your talk. The use of an interpreter will slow down your presentation so you may have time for only half as much material. Take this into account in planning your talk.

In addressing small groups you have a very wide choice of visual aids that you can use. Visitors in particular like to see something in addition to being spoken to. You should therefore arrange for a visual display even if you have time only to prepare a few flip-chart sheets.

Visitors often have to be shown round and this raises a number of particular problems. Decide in advance where you will stop and speak. Speaking on the move is not satisfactory. The visitors will be trailing behind and only those at the front will know what you are talking about. Having reached your first stopping place, turn and wait until your party arrives. You should not start to speak until the last to arrive has settled into his position in the group. Your opening 'Ladies and Gentlemen' will restore the degree of formality lost during the preceding stroll. It will also command silence. A lot can be learned by observing the technique used by professional guides in conducting parties around stately homes, show caves and such like. The guide, you will notice, knows exactly where to stand to make each presentation. He or she also knows exactly where the visitors are to stand. Features of interest that are to be seen in transit to the next stopping place are described in advance, not afterwards.

Speaking to small groups of people provides an excellent means of practice in public speaking. You should therefore welcome such opportunities and treat each one as a challenge. If you approach the event with the attitude that it is an unproductive chore that you could well do without, you will be missing a valuable chance to improve your public speaking skill.

20   Speaking at Committee Meetings

Committee meetings that scientists or engineers attend may be technical or non-technical. In technical committees, such as British Standards Institution committees, you will reasonably assume that the other members of the committee will have a similar background and level of knowledge as yourself. Preparation of your material will be eased: you will all be speaking the same language. In non-technical committees appropriate preparation of your material is extremely important. You may be speaking to executives responsible for finance, marketing, administration, etc. Such people will soon lose interest in technical detail and jargon. Your task is to prepare your material with your audience very much in mind.

A common situation is that you will have been asked to prepare a document for the meeting, and that you will be expected to speak at the meeting in order to introduce your document for consideration and discussion. Make sure that your document is prepared in sufficient time to allow its circulation before the meeting. Very long documents may need to be circulated weeks or even months before the meeting to allow detailed study by the members and their advisers. More usually the document will be short, less than about ten pages say, and the optimum time for prior study is probably about one or two weeks. Papers received before this will tend to be read and put on one side. The ideas will then lack an atmosphere of freshness when the matter comes up for discussion. Papers received less than a week before the meeting are in danger of not being given adequate attention prior to the meeting. At all costs avoid having to hand out copies of the document at the meeting, unless of course there is a particular reason why the matter must be dealt with in this way.

Your verbal presentation should be a guide to the document but not necessarily a summary of it. Members of the committee want to hear the essence of your message. What has to be decided? What are your conclusions and recommendations? How strongly do you feel about your recommendations? How clear-cut do you see the matter? A useful attitude of mind to adopt in preparing your material is to imagine that someone else has in fact written the document and that you have the task of presenting it. With this in mind you will not be tempted to extract large sections of the document verbatim. Rather you will view the document much as the other members of the committee will. You will ask what the document means and how it should be considered. Unlike the other members you will of course know the answers to such questions and your presentation will be based on providing the answers.

It is unlikely that you will have been given a time limit to work to, but this does not mean that you should speak at length. The briefer you are, the more your contribution will be appreciated.

Armed with your prepared notes, arrive early for the meeting. This will give you the opportunity to study the room and choose your seat. It is advantageous to be seated where the chairman can readily see you. If there is a telephone in the room, sit well away from it. Let someone else deal with interruptions. Similarly avoid sitting near the door. Otherwise you could find yourself involved in various tasks from issuing cups of coffee to re-directing someone who cannot find the right room. It is better to face the windows than to have your back to them. In this way your listeners will find it more comfortable looking at you as you speak, and you will more easily hold their attention. Avoid sitting in corner positions. Not only will you be less cramped in terms of table space but your presentation will seem more authoritative. There is a psychological effect on the audience that causes corner speakers to appear to be defensive and less certain of their statements.

From an examination of the agenda decide whether you will have an opportunity of saying something before you make your prepared presentation. The advantage of an earlier contribution is that it allows you to try out your voice, and feel the resonance of the occupied room. In addition it allows the other members to hear your voice, mentally accept it and associate it with your face. When the time comes for your presentation you will already have established a degree of acceptability. Obviously, your earlier contribution must be made on proper grounds. Any impression that it has been contrived will damage your credibility rather than improve it.

As you wait for your item on the agenda to arrive, pay attention to what is going on. You could be asked unexpectedly by the chairman for a comment or a piece of information. Even if this is unlikely you must still appear to the other members to be interested in what they are saying. They will be forming an impression of you before they hear your presentation. If they observe that you are bored or, worse still, on the verge of sleep, they are going to be less inclined to give you their full attention when it is your turn to speak. You can win allies in your yet-to-be-revealed cause merely by being interested.

Be formal in your presentation and in the subsequent discussion, even if the others are not. 'Mr Chairman, Ladies and Gentlemen' is your opening, or, of course, 'Mr Chairman, Gentlemen' if there are no ladies present. All your remarks should be addressed to 'Mr Chairman' and you should refer to the other members as 'Mr . . .' or 'Dr . . .'. The benefits of maintaining formality have been mentioned several times previously but in the meeting situation there is a further advantage of adopting these procedures. If there is a matter of disagreement the use of the procedures has the effect of preventing a personal feeling being attached to criticisms. In other words you can tell the chairman that you think Mr Brown is quite wrong, and this will not affect your personal relationship with John Brown anything like as much as if you were to say 'John, I think you are quite wrong'.

Your presentation should proceed as described in Chapter 5 but you will of course not stand but remain seated. Note that in a sitting position it is very easy to forget about the need for visual communication. Hold your notes in your hand rather than leave them on the table in front of you. This will prevent you sitting hunched with your head bent forward and eyes focused on the table. When it comes to question and discussion time you will be in an impromptu speaking situation. Follow the guidance given in Chapter 6.

Do not be in too much of a hurry to answer questions or criticisms. You will not give your best possible answer if you respond immediately, particularly if you are annoyed or under pressure. Ways of creating thinking time for general use are described in Chapter 6 but in a meeting you have other means at your disposal. You can wait until the chairman invites you to reply, making valuable use of the time that this takes. A further advantage of waiting is that someone else may be eager to comment, and the chairman, seeing that you are not jumping in with a response, may invite this further contribution. You can, alternatively, actively create such a situation by suggesting to the chairman that, before answering, it might be useful to invite others to add their views. Even if the invitation is not accepted you will have created several seconds of thinking time.

It hardly needs saying that you should not attempt to answer what you cannot answer. A simple statement to the effect that you do not know the answer takes less time than a wordy ramble round the subject and does more for your credibility and prestige. You should of course propose a way out. You might, for example, offer to supply information after the meeting, or to undertake an action subject to the point being confirmed.

Your final task is to ensure that you are clear as to the outcome of the consideration of your presentation. What actions have been agreed for yourself and for the others and when are they to be completed? If you are not clear as to the position or if you suspect that others are not, then ask the chairman for clarification.

21   Chairing Committee Meetings

The rules of procedure covering meetings are based on English parliamentary procedures and in detail are complex.[26,33] In that a chairman needs to be familiar with the rules of procedure and should apply them effectively, his job is not an easy one. However, most of the meetings that the average scientist or engineer is concerned with are not held rigidly within these rules. Motions are rarely put in a formal way and most decisions are reached without a formal vote.

Formal rules of procedure are more desirable when the meeting is large or concerned with major issues, and when those attending are not well acquainted with each other or are sharply divided in opinion. The rules tend to be relaxed when these conditions do not apply. Meetings chaired by scientists or engineers are not likely to involve differences of opinion so widely separated as might be so in some areas of commercial activity. The meetings will frequently be small and attended by people known to each other.

This chapter therefore is concerned with the chairman's need to be effective rather than his need to apply formal rules of procedure. Nevertheless, the chairman should take the precaution of making himself familiar with the formal rules and of arming himself with a suitable reference book. It can always happen that someone attending the meeting will insist on a matter being dealt with in a formal way. Under such circumstances the chairman should comply with the request.

In case it appears that we have totally abandoned formality let us restate once again our position on formality. Every public speaking situation requires a degree of formality and the degree to be aimed for is slightly more than that which is considered necessary. It is easy for the speaker to lower the level of formality but difficult for him to raise it once he has started his presentation.

We shall assume that you are to chair a meeting of a technical committee. There will be a formal agenda circulated in advance. The secretary will produce minutes of the meeting, though these may be summarised notes rather than minutes in the formal sense. Everyone will address you as Mr Chairman and everyone will pay lip service to the rule that all statements must be addressed to the chairman.

It will come as no surprise to you to see that my first piece of advice is to start the meeting on time. The advice is obvious but the need to give it bears some examination. Everyone agrees that meetings should start on time so why don't they? As chairman you will intend to start on time, but as the moment approaches you will experience a new feeling. A desire to give people 'a couple of minutes' will grow in your mind and, rather than sitting in silence, waiting, you will start a conversation with your neighbour to fill in a few moments. If you recognise the true origin of this feeling then you can escape its clutches. In reality your desire is not to give the others 'a couple of minutes' but to give yourself 'a couple of minutes'. The start of a meeting is a tense moment for you as chairman and your delay is a subconscious protective instinct. In most public speaking situations your start is signalled for you, but as chairman you must start without a signal. It is a good idea to look at your watch and start a mental countdown so that you start speaking on a precisely predetermined second of time. In this way not only will you start on time but your opening will be clear and authoritative and will catch attention. It is fatal to intend to start and then hover, looking round at inattentive faces and listening for lulls in the murmuring which will allow you to be heard. Such an approach results in a vague ill-defined opening.

Open the meeting in a positive manner. Welcome those attending and state what the meeting is: it is always possible that someone has found his way into the wrong room. Ensure that everyone present knows who everyone else is. This can conveniently be done by asking each in turn to introduce himself or herself. Before considering the agenda items give some indication of your plans. At what time do you intend to close the meeting? Is this to be a rigid arrangement? If so, do you intend to adjourn and continue the meeting at some later date, or are you going to insist on all matters being completed within the allocated time? If the meeting extends into the afternoon at what time will a break be taken for lunch and what are the arrangements for lunch? Has anyone any problems with regard to travelling plans? These questions appear obvious when written out in this fashion but chairmen infrequently give proper attention to them. The result is that everyone at the meeting feels slightly insecure and gives less than full attention to the matters under discussion. One is wondering if he will be home in time to take his children to the Cubs meeting. Another is considering his chances of catching the 4.35 train. Several have decided well before the end of the meeting to make no further contribution, in the hope thereby of speeding up progress. By indicating the overall plan at the outset, the chairman creates a comfortable atmosphere and establishes the desired sense of haste or, alternatively, a more leisurely pace.

As you introduce each agenda item treat the occasion as a separate presentation and prepare for it accordingly. Do not assume that because you know what the item is about you will automatically introduce it well. It is your job firstly to catch the attention, to reawaken tiring minds and raise enthusiasm. Find an opening remark that meets these needs. Your second task is to make clear what has to be decided. In order to do this you may have to review certain facts, policies or previous decisions. Good preparation will allow you to set the scene as concisely as possible.

You are now in a position to invite contributions, some prepared perhaps, some not, from the members around the table. In formal rules of procedure the impartiality of the chairman is stressed. In most business or technical committee meetings however the chairman will certainly not be unbiased. The chairman is frequently the most senior person present or the most expert in the matters under discussion. In your position as chairman, therefore, you will find that you have views on the decisions that you would like to see reached.

Your preparation for the meeting should take this into account. Your aim should be to guide the meeting towards the decisions you want without imposing your decisions or even appearing to impose your decisions.

The best way to get an idea accepted is to let your opponents believe that they have thought of the idea. Put this to good use by building on what arises in discussion. Discussion throws up a variety of suggestions. If you had unlimited time you could simply wait until you heard your solution coming from the mouth of someone else. In fact you do not have unlimited time but you can quickly spot contributions that are heading in the right direction. A question from you to the meeting following such a contribution may then produce something nearer the mark, and so on.

Having a preconception of the decision you want is not the same as saying that you cannot or should not change your mind. You have not necessarily anticipated all the points that are raised, and you may not know of all facts or consequences.

Listen carefully to the discussion and be prepared to modify your views.

It may be of course that your tactics are unsuccessful, in which case state your view and invite direct comment on it.

It would be unusual in a technical meeting to have to put a matter to the vote. It is not within the instincts of a professional scientist to support a concept on the basis of such a random process. If views are so divided it suggests that the meeting has in fact no proper basis for making the decision. In a business meeting of course the situation could be quite different. Where policy is concerned, opinion and judgment play a proper part in decision making.

If it is evident that more information is needed before a proper decision can be reached, then identify how the information is to be obtained and defer the matter until a subsequent meeting.

At the conclusion of each agenda item ensure that everyone, especially the secretary, is clear as to what has been decided. If action is required identify who is to do what and when it is to be done by.

At the end of the meeting thank those who have attended and highlight in a few words what has been accomplished. Each member should leave the meeting with a feeling that it has been worthwhile and that he is happy to continue to contribute to the work of the committee.

This short guide to chairmanship will not correspond with much of the chairmanship that you have seen in practice. But how do you rate the chairmen that you encounter in your business life? Not very highly I would guess. The 'weak' chairman is commonly encountered and not greatly respected. His meetings are of indeterminate length, discussion ranges far and wide and many items are held over until the next meeting. The 'strong' chairman is looked on more kindly. At least there are clear decisions, but is he using the committee meeting in the way that was intended or is he merely pulling rank and using it to transmit his instructions?

22 Informal Meetings

The situation to be considered in this final chapter is probably the least formal public speaking occasion in which you are involved. A small meeting with up to perhaps five or six people attending is a common event in all kinds of business. As a scientist or engineer it is likely that you are involved with such meetings regularly; so regularly in fact that you have probably never classed them as public speaking occasions.

To have failed to recognise the true nature of such meetings is to have missed opportunities in two respects. Firstly you have missed the opportunity of developing your public speaking skills. Secondly you have almost certainly not gained the maximum advantage from the meeting in relation to your purpose in attending. If you consider such meetings to be nothing more than conversation, albeit technical conversation, then you will treat them as such. Perhaps on reflection you have wondered why such meetings can be so unproductive, no more productive in fact than an average conversation.

The meeting will have an air of informality. There will be no chairman, so defined, but someone will have initiated the meeting and he will probably play a leading role. Everyone will probably be on first-name terms if it is a meeting internal to your organisation, but if more than one organisation is involved titles and surnames may be used. No minutes will be taken, each person relying essentially on his own notes. There may be an agreement that someone should produce a record of decisions or actions.

In spite of the informal atmosphere you should aim to establish a degree of formality when you speak. Most of the instruction given in Chapters 5 and 6 applies, with some obvious exceptions. You will speak while seated and the volume in your voice need not be more than a normal speaking level.

Most of what you say will be impromptu, but this does not mean that preparation is not important. Preparation is all the more important exactly because you do not know precisely what you will need to say.

Start your preparation by establishing your aims. What would you like to see achieved as a result of the meeting? What is the minimum that would satisfy you? What would be the consequences of not achieving the minimum? You may be thinking by now that we are over-rating this event. After all, you attend many such meetings merely to give technical advice or to gain information from a colleague or other organisation. Nevertheless, you have a purpose even if it relates largely to what you do not want to result from the meeting. You may be attending to present some technical data, but you have no wish to find at the end of the meeting that you have committed your department to six months further experimentation when you cannot spare the necessary resources. Consider your aims in a negative as well as a positive sense.

The next task is to prepare notes. There will be a number of contributions that you intend to make at appropriate stages in the meeting. Use separate cards for each of these contributions. In this way you make your plan flexible and can readily raise the right point at the right time. Some of your intended contributions may be conditional on what others say. More on this matter later, but for now let us recognise that you can usefully prepare alternative notes to cover the same contribution, and you can usefully prepare some notes that may not be required at all. Inevitably, in such a meeting you will not be able to prepare in advance all the notes you may need. There will be a number of grey areas in which it is not clear what points will be raised by others or how the discussion will go. Attempt to cover these areas by use of brain patterns as described in Chapter 6. A skeleton brain pattern prepared in advance can be added to as the meeting proceeds. You can then extract notes from the pattern during the meeting as required, or speak using appropriate parts of the pattern as your notes.

In preparing what you intend to say it is important to consider the other people who will be attending the meeting.

If you know them individually consider the attitudes each is likely to adopt. Which are the bright ones? Which are the sheep? Who is likely to dispute everything you say? Who can you rely on for support? If the people are unknown to you, you may simply have to rely on your expectation of their department's view or that of their company. But you may be able to do better than this. Can you make enquiries in advance that might give you a guide to their personalities and attitudes?

Having recognised the attributes of each of the people you will join in discussion, reflect on the fact that without exception each has his pride. If anyone has his pride hurt as a result of what you say, you may win a victory in the short term but not necessarily in the long term. That is not to say that you cannot be critical. People will accept criticism, will accept being proved wrong, provided they can retain their pride. Each person attending a meeting wants to take something away with him. You should ask yourself what it is that each wants. How can your aims be seen to give something that each wants? You may not be able to satisfy all requirements but it is usually possible with foresight to ensure that no one, because of what you say, leaves the meeting feeling that he has lost his self respect and has got nothing out of the meeting.

If it is a meeting that you are arranging then you will be able to decide at what time it should start. Conventionally you will start on the hour or at half-past the hour. But by arranging the time for ten minutes past or five minutes to, say, you can make it clear that in your view time is valuable and that you intend the meeting to proceed in a businesslike fashion without delays.

If you have any say in the seating arrangement remember that a physical alignment of groups of people can emphasise their differences of opinion. We have all attended meetings in an office where the occupant sits behind his desk, back to the wall, while the rest of us face him in an arc or worse still a straight line. Psychologists will know why offices tend to be arranged in this way but the arrangement is not conducive to productive meetings. The desk presents a barrier between the leader and the others, each of the latter gaining strength from the physical alignment with his neighbours. It is much better to sit round a table and to split up people who would be expected to share similar views.

Meetings tend to proceed at the pace set in the first few minutes. If it is your meeting, therefore, your opening contribution is one of your most important. If you insist on leisurely recapping the history of the subject to be discussed, then each following speaker will feel happy to continue in the same vein. On the other hand, if you start by stating clearly and concisely what the questions are that have to be resolved, everyone will be reluctant to stray from the subject.

Even if you have prepared well, you cannot be sure that you will accomplish your aims. It will become evident during the discussion that you have some opposition. To a degree you may have anticipated this and prepared your answers, but there may develop a situation in which you will either lose your point or win it by producing an impromptu winning tactic. Some possible tactics are listed below.

(i) See if you can get someone to make a commitment related to your proposals. He may dispute your figures or your background information. Ask him if he can supply data for a subsequent meeting. He is likely to agree and by the next meeting he has already aligned himself with you to a degree by his involvement. There may still be differences of course but you have improved your position.

(ii) Instead of pushing your own view too hard, concentrate on the opposing view. Analyse it and explore it until the opposition begin to see difficulties and begin to divide. Let them talk themselves out of it.

(iii) Watch carefully for the right time to make your contributions. Judge the atmosphere. When a suggestion is received enthusiastically the time is not right to raise a doubt. If frustration and deadlock prevail a new approach will be welcomed. If your idea is accepted, resist pushing home your victory to the point at which resentment or doubt sets in.

(iv) Make sure that the opposing argument is what it appears to be. Perhaps your contestant is grinding his axe. Perhaps he needs a victory more than he needs acceptance of his argument. You may be able to allow him what he needs without losing the debate.

(v) An approach that brings hope to a meeting in difficulties, is often welcomed with open arms. Obviously you have to be sure that your idea will live up to its promise.

(vi) Remember that decisions are often taken in favour of the least disadvantageous of the possibilities rather than the most advantageous.

(vii) Presumption can often win the day. If you are prepared to risk adopting the view that you have a measure of agreement, you have something to build on. If you can build fast enough your dubious foundation may be overlooked.

(viii) Is there anything you can offer to undertake to do as a condition of acceptance of your scheme? Everyone has plenty to attend to, and will breathe a sigh of relief at the thought of follow-up action falling to someone else.

(ix) Proposals that appeal to a sense of justice and fairplay are often accepted: hence the popularity of the compromise solution. Many one-sided solutions, however, can be presented so as to appear fair to all.

(x) A touch of humour can often be introduced to draw attention away from a dispute and create a better atmosphere for its solution.

If in spite of application of such methods, you find that you have to lose a point, then do so without covering up the fact. Make the opposition win the point. Give them their victory. This will give you advantage in subsequent encounters. In losing your point, however, preserve your aims as far as possible. Ask for time to give further consideration, to obtain more information, or to take advice.

It is important to recognise that discussion cannot reconcile views that are fundamentally different. Discussion is often overrated as a means of solving problems. Many meetings are called in the mistaken belief that if everyone is allowed his say, the right answer will emerge. A meeting is productive when those attending have some common objective. Each may have a different opinion about the best way to achieve the objective but at least there is an agreed aim. Opinions differ because each has his personal aims in addition to the common aim. Under such circumstances, discussion is worthwhile. Subsidiary targets are exposed, weighed against each other and accepted or rejected. Points are won and lost and progress is made towards the desired conclusion. On the other hand if a common objective is lacking, the meeting is likely to waste everyone's time. The aims of individuals will still be examined, points will still be won and lost, but measurable progress will be hard to find.

Finally, a few words on ending your meeting. The way you end is important because you want each person to leave feeling that it has been worthwhile and keen to carry out the follow-up actions. Stress the fact that things have been accomplished. List the specific conclusions and list the actions to be taken. Avoid being drawn into a further difficulty that someone has just spotted as the meeting is breaking up. Offer to deal with the matter after the meeting if at all possible.

Epilogue: a Few Words to the Members of the Audience

The preceding chapters have been written in the belief that there is considerable room, and a desire, for improvement in the standard of public speaking among scientists and engineers. The advice given has been directed to those speakers who wish to contribute to a raising of the standard.

In these closing paragraphs I would like to address members of the audience. We are all members of the audience at one time or another, and it is in this capacity that we suffer from poor presentations. Great improvement is unlikely to take place until audiences become more critical. Attending conferences is an expensive activity and we have a right to expect value for money, just as we would if attending a theatrical performance. Meetings also are expensive, though perhaps less obviously so. A glance round the table, an estimate of salaries and overheads, and a little mental arithmetic will soon confirm the fact.

Perhaps we should be prepared to give voice to our criticisms. We could complain to conference organisers as we would if, for example, the catering arrangements were unsatisfactory. In reviewing conferences for journals we could comment not just on the quality of the papers but on the quality of presentations. We could identify speakers who were particularly effective, so that conference organisers would start to plan in terms of 'who can speak well?' rather than simply 'who is the technical expert?'

Within our own organisations we could encourage a better attitude. Young scientists need training in public speaking and need to see that skill in this field will be given due recognition.

When we send them to present views or reports at meetings we have a duty to see that they are suitably equipped for the task. We should consider it part of our job to be responsible for their verbal work just as we are for their written work.

Somehow we need to overcome the sensitivity attached to criticism of public speaking. This delicacy is by no means confined to the technical world. The Guild of Professional Toastmasters elects the most boring speaker of the year but, for professional reasons, will not publicise the winner's name until AD 2000! How can we remove the taboo that inhibits us from complaining and insisting on improvements? Who will be the first to declare: 'This verbal presentation we are being subjected to is not of the quality we have a right to expect?'

Appendices

Appendix 1   References and Further Reading

1   Commentary *Electronic Components* September 1969 p 965.
2   *Sunday Express* 13 January 1974.
3   Mills H R 1977 *Techniques of Technical Training* 3rd edn (London: Macmillan).
4   Buzan T 1974 *Use Your Head* (London: British Broadcasting Corporation Publications Department).
    . Surprisingly, this book is not aimed at the public speaker, nor is public speaking mentioned in the book. Nevertheless I consider the book to be one of the most useful aids that a public speaker can have.
5   Powell L S 1978 *A Guide to the Use of Visual Aids* (London: British Association for Commercial and Industrial Education). A short but extremely useful book concerned not only with using aids but with making them. Many references are included.
6   Seekings D 1981 *How to Organise Effective Conferences and Meetings* (London: Kogan Page).
7   Powell L S 1980 *A Guide to the Overhead Projector* (London: British Association for Commercial and Industrial Education).
8   British Standards Institution 1977 Preparation of artwork for lecture slides *PD 6482: 1977*.
9   Godard P 1979 Editorial *Materials Performance* **18** (12) 7. The theme of this editorial is a complaint about the standard of slides used at NACE (National Association of Corrosion Engineers) Conferences.
10 Lawrence R S 1964 *A Guide to Public Speaking* (London: Pan) pp 136–8. This is a useful, though not very well organised, book covering all aspects of public speaking. It is well worth reading, particularly by those intending to take examinations, as the author is an examiner for the London Academy of Music and Dramatic Art.
11 Henderson A M 1956 *Good Speaking* (London: Pan) pp 14–20. This excellent book is unfortunately out of print at the time of writing. It does not cover all aspects of public speaking but concentrates on diction and voice development.
12 *The Oxford Dictionary of Quotations* 3rd edn 1979 (London: Oxford University Press).

13 *The Concise Oxford Dictionary of Quotations* 1964 (London: Oxford University Press).

14 Cohen J M and Cohen M J 1960 *The Penguin Dictionary of Quotations* (Harmondsworth: Penguin).

15 Cohen J M and Cohen M J 1971 *The Penguin Dictionary of Modern Quotations* (Harmondsworth: Penguin).

16 Mackay A L 1977 *The Harvest of a Quiet Eye: A Selection of Scientific Quotations* (Bristol: The Institute of Physics).

17 Wilson F P 1970 *Oxford Dictionary of English Proverbs* 3rd edn (London: Oxford University Press).

18 McWhirter N *Guinness Book of Records* (London: Guinness Superlatives) published annually.

19 Evans I H 1970 *Brewer's Dictionary of Phrase and Fable* (London: Cassell).

20 Boyd L M 1979 *Boyd's Book of Odd Facts* (New York: Sterling).

21 Ward P 1980 *Dictionary of Common Fallacies* vols 1 and 2 (London: Oleander).

22 Weber R L 1973 *A Random Walk in Science* (Bristol: The Institute of Physics).
    A remarkable collection of extracts, stories, quotations, paradoxes, etc, some amusing some serious, from various branches of science.

23 Lenihan J and Fleming J B 1979 *Science in Action* (Bristol: The Institute of Physics).
    A collection of articles written for the layman dealing with unusual aspects of science.

24 Walker J 1975 *The Flying Circus of Physics* (Chichester: Wiley).
    Problems in physics relating to common experience that are surprisingly difficult to solve. This book is designed to challenge, and deflate, any scientist.

25 Walker J 1977 *The Flying Circus of Physics with Answers* (Chichester: Wiley).
    As reference 24, but for busy scientists and for those who gave up struggling with reference 24 after two years.

26 Gondin W R and Mammen E W 1970 *The Art of Speaking Made Simple* (London: W H Allen).
    A well organised book covering not only public speaking but also conversation and interviews. The book is useful for reference but the advice given is peculiar in some instances.

27 Bullard A M 1960 *Improve Your Speech* (London: Anthony Blond).

28 Parkin K 1969 *Ideal Voice and Speech Training: A Book of Exercises* (London: French).

29 Parkin K 1969 *Anthology of British Tongue-twisters* (London: French).
30 Kirby P L 1977 Cross-Channel communication *Physics Bulletin* **28** (1) 7–8.
31 Boyce J F and Michelson D 1977 Letters *Physics Bulletin* **28** (4) 150.
32 Hird B 1977 Letter *Physics Bulletin* **28** (5) 199.
33 Lord Citrine 1952 *ABC of Chairmanship* (London: NCLC Publishing Society).

Appendix 2   Preparation checklists

STAGE I: PREPARATION AS SOON AS POSSIBLE (MONTHS BEFORE)

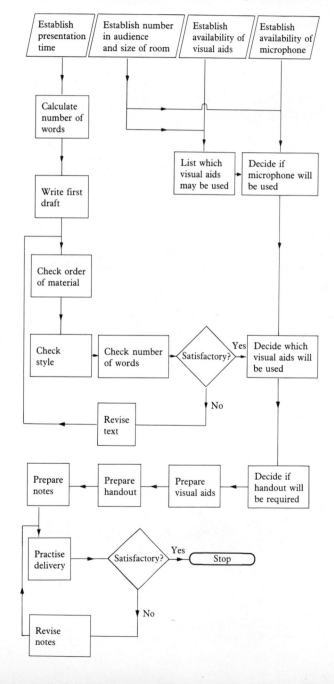

STAGE II: PREPARATION BEFORE THE MEETING (HOURS BEFORE)

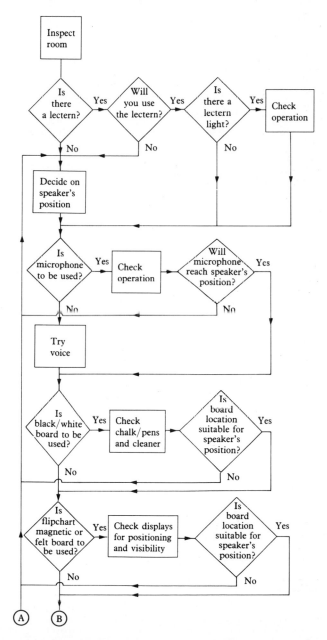

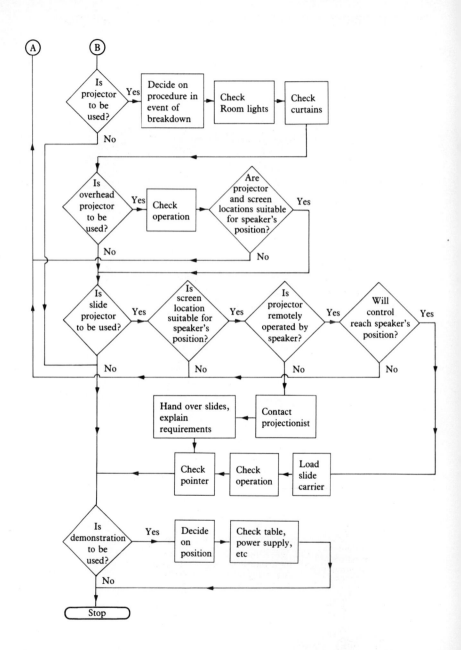

STAGES III AND IV: FINAL PREPARATIONS

Stage III (minutes before)   *Stage IV (seconds before)*

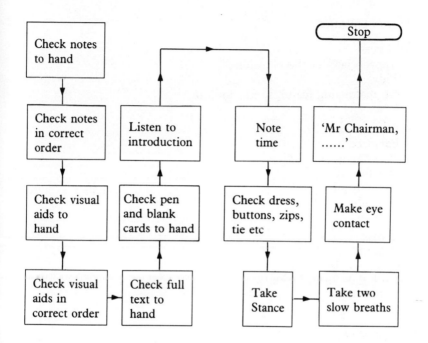

Appendix 3   Performance Evaluation Checklist

1   APPEARANCE

(a) Dress
Appropriate for the occasion
Tidy
No distracting features, e.g. loud tie

(b) Stance
Balanced
No swaying or rocking
Not too stiff
Not slovenly
Not fidgety

(c) Poise
Relaxed
Comfortable
No indications of tension

2   DELIVERY

(a) Visual Communication
Maintained
Definite
Directed to all parts of audience

(b) Gestures
Meaningful
Restricted in use

(c) Mannerisms
Absent

(d) Use of Visual Aids
Familiarity with controls
Smooth introduction of material
No loss of audience contact

(e) Use of Microphone
  Familiarity with controls
  No tapping or blowing, etc
  No loss of sound due to head movement

(f) Use of Notes
  Full text not read
  Notes not fumbled
  Notes used only when required
  No serious departures from notes
  Notes held sufficiently high
  No attempt to hide notes

(g) Use of Words
  Fluent
  No unfinished sentences
  No construction changes in mid-sentence
  No false starts

(h) Timing
  As allocated
  Not rushed at end

3 DICTION

(a) Articulation
  Clear
  Not slurred or drawled
  Not overprecise
  Adequate movement of jaw, lips, tongue

(b) Volume
  Adequate
  Not 'breathy'
  No fading at end of sentences
  Variations used effectively

(c) Emphasis
  Used effectively
  Stress level sufficient
  No false use
  No excessive use

(d) Pace
    Not too fast or slow
    Variations used effectively

(e) Pause
    Correct length
    Not lacking in use
    Not used excessively
    Suitable allowance for laughter

(f) Phrasing
    Natural
    Not jerky
    Awareness of meaning

(g) Pitch
    Not too high or low
    Variations used effectively
    No nasal tone

4   MATERIAL

(a) Vocabulary
    Suitable for audience
    Simple where possible
    No clichés
    No jargon
    No repetitive use of particular words

(b) Word Groupings
    Short sentences or phrases
    Simple constructions

(c) Structure
    Recognisable introduction
    Recognisable body
    Recognisable conclusion

(d) Introduction
    Opening caught the attention
    Unusual opening

Established link with audience
Set the scene adequately
Brief

(e) Body
Flowed logically
Main points clear
Main points linked
Not too much information
Not too much minor detail
Sufficient periodic recapitulation

(f) Conclusion
Adequate summary of points
Ending signalled
Final message clear
Final message memorable
Link with opening
Brief
No false endings

5   OVERALL IMPACT

Met level of knowledge of audience
Met need for knowledge of audience
Extent of material adequate
Depth of treatment adequate
Purpose achieved
Enjoyed by audience
Audience left wanting more

Appendix 4   Public Speaking Examinations

Key: P, prepared talk or speech; I, impromptu talk or speech; R, reading; M, memorised presentation; O, oral précis; D, discussion.

GUILDHALL SCHOOL OF MUSIC AND DRAMA
BARBICAN, LONDON, EC2Y 8DT.   *071- 628-*
                                       2571

Spoken English

Introductory	P		R	D
Grade 1	P		R	D
Grade 2	P		R	D
Grade 3	P		R	D
Grade 4	P		R	D
Grade 5	P		R	D
Grade 6	P(2)		R	D
Grade 7	P	I	R	D

Certificate in Speaking
P   I   R   M   D and written paper

Teacher's Diploma in Public Speaking
P   I   R   M   D and written paper

Examination Centres
England   (67 centres)
Wales   (2 centres)
Scotland   (6 centres)
Northern Ireland   (4 centres)
Channel Islands
Eire   (2 centres)
Australia
Hong Kong   *071-373-9883*

LONDON ACADEMY OF MUSIC AND DRAMATIC ART
TOWER HOUSE, 226 CROMWELL ROAD, LONDON, SW5 0SR.

Public Speaking

Bronze Medal	P(2)	I	
Silver Medal	P(2)	I	
Gold Medal	P(2)	I	R D

Associate Public Speaking Diploma (ALAM)
P(2) I R D

Examination Centres
England (40 centres)
Wales (4 centres)
Scotland (4 centres)
Northern Ireland
Eire

LONDON COLLEGE OF MUSIC
GREAT MARLBOROUGH STREET, LONDON, W1V 2AS.

Oral Communication

Grade I	P		R	D
Grade II	P		R	D
Grade III	P		R	D
Grade IV	P		R	D
Grade V	P		R	D
Grade VI	P(2)	I	R	D
Grade VII	P(2)	I	R	D
Grade VIII	P(2)	I	R	D

Diploma of Associate in Public Speaking (ALCM)
P(2) I R D and written paper

Diploma of Licentiate in Public Speaking (LLCM)
P(3) I R D and written paper

Diploma of Fellowship in Public Speaking (FLCM)
P(4) I

Examination Centres
England (84 centres)
Wales (17 centres)
Scotland (6 centres)
Northern Ireland (9 centres)
Eire (7 centres)
Malta

TRINITY COLLEGE OF MUSIC
MANDEVILLE PLACE, LONDON, W1M 6AQ.

Effective Speaking

Initial		R	M	D		
Grade	I	P	I	R	M	
Grade	II	P	R	M	D	
Grade	III	P	R	M	D	
Grade	IV	P	R	M	O	D
Grade	V	P	I	R	M	D
Grade	VI	P	I	R	M	D
Grade	VII	P	I	R	M	D
Grade	VIII	P	I	R	M	D
Grade Adult	IV	P	I	R(2)		D
Grade Adult	V	P	I	R(2)		D
Grade Adult	VI	P	I	R(2)		D
Grade Adult	VII	P	I	R(2)		D
Grade Adult	VIII	P	I	R(2)		D

Licentiate Performer's Diploma in Effective Speaking (LTCL)
P   I(2)   R(2)   O   D and written paper

Examination Centres
The College has about 1000 examination centres at which effective speaking examinations could be taken depending on demand. In practice such examinations are taken regularly at the following centres:

England   (11 centres)
Eire
Australia   (6 centres)
Canada   (3 centres)
India
New Zealand   (8 centres)
Singapore
Sri Lanka

Appendix 5 Some Examples of Material from Unusual Sources

1 MATHEMATICS AND MENSURATION

Nursery rhymes
How many miles to Babylon?
And he walked a crooked mile

Proverbs and sayings
The Eternal triangle
A square peg in a round hole
At sixes and sevens
Give him an inch and he'll take a yard
To put two and two together
He knows how many beans make five
Look after number one
Measure thrice before you cut once
Six of one and half a dozen of the other
Safety in numbers

Biblical
Be fruitful and multiply *Gen.* 1 : 28
Multiply my signs *Exod.* 7 : 3
Found out my riddle *Judg.* 14 : 18
Like the deaf adder *Ps.* 58 : 4
Threescore years and ten *Ps.* 90 : 9
Teach us to number our days *Ps.* 90 : 12
So they cast lots *Jonah* 1 : 7
Can add one cubit *Matt.* 6:27
Until seventy times seven *Matt.* 18:22
With what measure ye mete, it shall be measured to you *Mark* 4 : 24
Ninety nine just persons *Luke* 15 : 7
Which no man could number *Rev.* 7 : 9
His number is six hundred threescore and six *Rev.* 13 : 18

Shakespeare
Play at dice *The Merchant of Venice* II, 1
Shrunk to this little measure *Julius Caesar* III, 1

Purchased at an infinite rate *The Merry Wives of Windsor* II, 2
Divinity in odd numbers *The Merry Wives of Windsor* V, 1
Divide a minute into a thousand parts *As You Like It* IV, 1
The trick of singularity *Twelfth Night* II, 5
Out, hyperbolical fiend! *Twelfth Night* IV, 2
The baby figure of the giant mass *Troilus and Cressida* I, 3
Zed! thou unnecessary letter! *King Lear* II, 2

2 ASTRONOMY AND GEOLOGY

Nursery rhymes
The cow jumped over the moon
The moon's in a fit
Twinkle, twinkle little star

Proverbs and Sayings
If the mountain will not go to Mahomet
Patience wears out stones
Stars are not seen by sunshine
Still waters run deep
Time and tide wait for no man
The man in the moon
To make a mountain out of a mole-hill

Biblical
In the beginning *Gen.* 1 : 1
The stars in their courses *Judg.* 5 : 20
The morning stars sang together *Job* 38 : 7
Before the mountains were brought forth *Ps.* 90 : 2
The sun shall not smite thee by day nor the moon by night *Ps.* 121 : 5
All the rivers run into the sea; yet the sea is not full *Eccles.* 1 : 7
There is no new thing under the sun *Eccles.* 1 : 8
For the eyes to behold the sun *Eccles.* 11 : 7
Fair as the moon, clear as the sun *S. of S.* 6 : 10
The new moons *Isa.* 1 : 13
The desert shall rejoice, and blossom as the rose *Isa.* 35 : 1
Millstone *Matt.* 18 : 6
The stones would immediately cry out *Luke* 19 : 40
One star different from another star in glory *1 Cor.* 15 : 41
Let not the sun go down upon your wrath *Eph.* 4 : 26
Wandering stars, to whom is reserved the blackness of darkness forever *Jude* 13

Shakespeare
Give a name to every fixed star *Love's Labour's Lost* I, 1
The worshipped sun *Romeo and Juliet* I, 1
The inconstant moon *Romeo and Juliet* II, 2
As boundless as the sea *Romeo and Juliet* II, 2
The everlasting flint *Romeo and Juliet* II, 6
I'll put a girdle round about the earth *A Midsummer Night's Dream* II, 1
The burnished sun *The Merchant of Venice* II, 1
Born under a rhyming planet *Much Ado About Nothing* V, 2
There are no comets seen *Julius Caesar* II, 2
The Mars of malcontents *The Merry Wives of Windsor* I, 3
The lazy foot of time *As You Like It* III, 2
Like stars, start from their spheres *Hamlet* I, 5
Doubt thou the stars are fire, doubt that the sun doth move *Hamlet* II, 2
King of infinite space *Hamlet* II, 2
A great while ago the world begun *Twelfth Night* V, 1
A bright particular star *All's Well That Ends Well* I, 1
Th' inaudible and noiseless foot of time *All's Well That Ends Well* V, 3
Deserts idle, rough quarries, rocks and hills *Othello* I, 3
A huge eclipse of the sun and moon *Othello* V, 2
The earth hath bubbles *Macbeth* I, 3
The unnumbered idle pebbles *King Lear* IV, 6
The demi-Atlas of this earth *Antony and Cleopatra* I, 5
The wat'ry star *The Winter's Tale* I, 2
Of his bones are coral made *The Tempest* I, 2
Like a shooting star *King Richard II* II, 4
The pale-faced moon *King Henry IV Part 1* I, 3
Two stars keep not their motion in one sphere *King Henry IV Part 1* V, 4
O polished perturbation! *King Henry IV Part 2* IV, 4

3 METEOROLOGY

Nursery Rhymes
When the wind blows the cradle will rock
The north wind doth blow
Rain, rain go away

Proverbs and Sayings

Every cloud has a silver lining
February fill dyke
He knows which way the wind blows
It is an ill wind that blows nobody any good
It never rains but it pours
Like a dying duck in a thunderstorm
Rain before seven, fine before eleven
When the wind is in the east
Red sky at night, shepherd's delight

Biblical

A little cloud out of the sea *1 Kgs.* 18 : 44
The wings of the wind *Ps.* 18 : 10
Whiter than snow *Ps.* 51 : 7
Let the floods clap their hands *Ps.* 98:8
The wind passeth over it *Ps.* 103 : 15
In a very rainy day *Prov.* 27 : 15
Observeth the wind . . . regardeth the clouds *Eccles.* 11 : 4
The winter is past, the rain is over and gone *S. of S.* 2 : 10
Neither can the floods drown it *S. of S.* 8 : 7
As white as snow *Isa.* 1:18
They have sown the wind and they shall reap the whirlwind *Hos.*
8 : 7
His sun rise on the evil and on the good, and sendeth rain *Matt.*
5 : 45
The wind bloweth where it listeth *John* 3 : 8

Shakespeare

More inconstant than the wind *Romeo and Juliet* I, 4
Too like the lightning *Romeo and Juliet* II, 2
The idle wind, *Julius Caesar* IV, 3
Let the sky rain . . . let it thunder *The Merry Wives of Windsor* V, 5
Lusty winter, frosty, but kindly *As You Like It* II, 3
Blow, blow, thou winter wind *As You Like It* II, 7
For the rain it raineth every day *Twelfth Night* V, 1
In thunder, lightning, or in rain *Macbeth* I, 1
Chaste as the icicle *Coriolanus* V, 3
As free as the mountain winds *The Tempest* I, 2
Is not their climate foggy, raw, and dull? *King Henry V* III, 5
The winter of our discontent *King Richard III* I, 1
Comes a frost, a killing frost *King Henry VIII* III, 2

4 CHEMISTRY, MEDICINE, METALLURGY AND MATERIALS

Nursery Rhymes
Silver buckles at his knee
But a silver nutmeg and a golden pear
With silver bells

Proverbs and Sayings
A good piece of steel is worth a penny
All that glisters is not gold
Blood is thicker than water
Desparate diseases have desparate remedies
An apple a day keeps the doctor away
Strike while the iron's hot
Touch wood
You cannot get blood out of a stone
You cannot see the wood for the trees

Biblical
Unstable as water *Gen.* 49 : 4
The price of wisdom is above rubies *Job* 28 : 18
A rod of iron *Ps.* 2 : 9
Much fine gold *Ps.* 19 : 10
Fretting *Ps.* 39 : 12
Fetters of iron *Ps.* 149 : 5
Wisdom is better than rubies *Prov.* 8 : 11
As a jewel of gold *Prov.* 11 : 22
Iron sharpeneth iron *Prov.* 27 : 17
Her price is far above rubies *Prov.* 31 : 10
Precious ointment *Eccles.* 7 : 1
The ointment of the apothecary *Eccles.* 10 : 1
Beauty for ashes *Isa.* 61 : 3
Is there no balm . . . no physician *Jer.* 8 : 22
The image's head was of fine gold . . . silver . . . brass . . . iron . . .
clay *Dan.* 2 : 32
Ye are the salt of the earth *Matt.* 5 : 13
Where the moth and rust doth corrupt *Matt.* 6 : 19
Need not a physician *Matt.* 9 : 12
Thirty pieces of silver *Matt.* 26 : 15
Physician, heal thyself *Luke* 4 : 23
Silver and gold have I none *Acts* 3 : 6
Oppositions of science falsely so called *1 Tim.* 6 : 20
The substance of things hoped for *Heb.* 11 : 1

A rod of iron *Rev.* 2 : 27
And the leaves of the tree were for the healing of the nations *Rev.*
22 : 2
The medicine of life *Sir.* 6 : 16
He that toucheth pitch *Sir.* 13 : 1
Honour a physician *Sir.* 38 : 1

Shakespeare
He that is giddy *Taming of the Shrew* V, 2
Rust, rapier *Love's Labour's Lost* I, 2
Saint-seducing gold *Romeo and Juliet* I, 1
Drawn with a team of little atomics *Romeo and Juliet* I, 4
Whose blood is warm within *The Merchant of Venice* I, 1
Endure the toothache patiently *Much Ado About Nothing* V, 1
The elements so mixed in him *Julius Caesar* V, 5
Of unimproved metal hot and full *Hamlet* I, 1
Each petty artery in this body *Hamlet* I, 4
Here's metal more attractive *Hamlet* III, 2
My pulse . . . doth temperately keep time *Hamlet* III, 4
Diseases, desparate grown *Hamlet* IV, 3
No other medicine but only hope *Measure for Measure* III, 1
Ay, past all surgery *Othello* II, 3
Nor all the drowsy syrups of the world *Othello* III, 3
Their medicinal gum *Othello* V, 2
He that sleeps feels not the toothache *Cymbeline* V, 4
By medicine life may be prolonged *Cymbeline* V, 5
Nor brass, nor stone, nor parchment *The Winter's Tale* I, 2
Gilded loam, or painted clay *King Richard II* I, 1
This villainous saltpetre *King Henry IV, Part 1* I, 3
How has he the leisure to be sick *King Henry IV, Part 1* IV, 1

5   HEAT

Nursery Rhymes
Her coat is so warm
He'll sit in a barn and keep himself warm

Proverbs and Sayings
Fire is a good servant but a bad master
Many irons in the fire
There is no smoke without fire

Biblical

The bush burned with fire *Exod.* 3:2
As the sparks fly upwards *Job* 5:7
As smoke is driven away, . . . as wax melteth before the fire *Ps.* 68:1
His clothes not be burned *Prov.* 6:27
Coals of fire upon his head *Prov.* 25:22
I am warm, I have seen the fire *Isa.* 44:16
A burning fiery furnace *Dan.* 3:11
As a firebrand plucked out of the burning *Amos* 4:11
And heat of the day *Matt.* 20:12
And the fire is not quenched *Mark* 9:44
Cloven tongues like as of fire *Acts* 2:3
How great a matter a little fire kindleth *Jas.* 3:5
As a flame of fire *Rev.* 1:14
Lukewarm, and neither cold nor hot *Rev.* 3:16

Shakespeare

Doth burn the heart to cinders *Titus Andronicus* II, 1
Where two raging fires do meet *Taming of the Shrew* II, 1
Fire, that is closest kept *The Two Gentlemen of Verona* I, 2
Shows a hasty spark *Julius Caesar* IV, 3
My flame lacks oil *All's Well That Ends Well* I, 2
To hide the sparks of nature *Cymbeline* III, 3
Heat not a furnace . . . so hot that it do singe yourself *King Henry VIII* I, 1

6 LIGHT

Nursery Rhymes

Can I get there by candlelight?
Here comes a candle to light you to bed

Proverbs and Sayings

A crooked stick will have a crooked shadow
A straight stick is crooked in the water
Many hands make light work. (Think about it!)
Seeing is believing
The darkest hour is nearest the dawn
The eye is the mirror of the soul
The game is not worth the candle
Two blacks do not make a white
What the eye don't see the heart don't grieve for.

Biblical
Let there be light *Gen.* 1 : 3
Even darkness which may be felt *Exod.* 10 : 21
A lamp unto my feet *Ps.* 119 : 105
Is as the shining light *Prov.* 4 : 18
Light excelleth darkness *Eccles.* 2 : 13
Have seen a greater light *Isa.* 9 : 2
For thy light is come *Isa.* 60 : 1
Neither do men light a candle *Matt.* 5 : 15
To give light to them that sit in darkness *Luke* 1 : 79
And your lights burning *Luke* 12 : 35
The light shineth in darkness *John* 1 : 5
Men loveth darkness rather than light *John* 3 : 19
For now we see through a glass, darkly *1 Cor.* 13 : 11
I shall light a candle *2 Esd.* 14 : 25

Shakespeare
Light, seeking light, doth light of light beguile *Love's Labour's Lost*
I, 1
Little candle throws his beams *The Merchant of Venice* V, 1
The glow-warm shows *Hamlet* I, 5
To hold . . . the mirror up to nature *Hamlet* III, 2
As we with torches do *Measure for Measure* I, 1
Give me the ocular proof *Othello* III, 3
Put out the light, and then—put out the light *Othello* V, 2

7   SOUND

Nursery Rhymes
My master's lost his fiddling stick
Ding dong bell
The cat and the fiddle
And drummed them out of town
Come blow up your horn
And he called for his fiddlers three
Say the bells of St. Clements
And she shall have music wherever she goes
Tom, Tom the piper's son

Proverbs and Sayings
Creaking shoes are not paid for
Deaf as a post

Empty vessels make the most noise
In at one ear and out at the other
Silence is golden
As sound as a bell

Biblical
He saith among the trumpets *Job* 39 : 25
Play skilfully with a loud noise *Ps.* 33 : 3
A merry noise *Ps.* 47 : 5
The players on instruments followed after *Ps.* 68 : 25
We hanged our harps *Ps.* 137 : 1
The loud cymbals *Ps.* 150 : 5
Cornet, flute, harp, sackbut, psaltery, dulcimer *Dan.* 3 : 5
As sounding brass, or a tinkling cymbal *1 Cor.* 13 : 1

Shakespeare
Bright Apollo's lute, strung with his hair *Love's Labour's Lost* IV, 1
The care where Echo lies *Romeo and Juliet* II, 2
Concord of sweet sounds *The Merchant of Venice* V, 1
Silence is the perfect herald of joy *Much Ado About Nothing* II, 1
Jangled, out of tune and harsh *Hamlet* III, 1

8 MECHANICS

Nursery Rhymes
And threw him down the stairs
The cow jumped over the moon
Humpty Dumpty had a great fall
When the wind blows the cradle will rock
When the bough breaks the cradle will fall
Jack fell down and broke his crown
See-saw, Margery Daw
With my bow and arrow

Proverbs and Sayings
Better to bend than to break
Constant dropping wears away the stone
Give a fool rope enough and he will hang himself
Have two strings to your bow
The best end of the stick
Little strokes fell great oaks

One good turn deserves another
The higher up, the greater fall
The strength of a chain is its weakest link
The thin end of the wedge
Those who live in glass houses should not throw stones

Biblical
Hewers of wood and drawers of water *Josh.* 9:21
Canst thou draw out leviathan with an hook? *Job* 41:1
Dash them in pieces like a potter's vessel *Ps.* 2:9
The arrow that flieth by day *Ps.* 91:5
A threefold cord is not quickly broken *Eccles.* 4:12
The silver cord be loosed or the golden bowl be broken *Eccles.* 12:6
A drop of a bucket, and are counted as the small dust of the balance *Isa.* 40:15
A bruised reed shall he not break *Isa.* 42:3
Thou are weighed in the balance *Dan.* 5:27
The axe is laid unto the root of the trees *Matt.* 3:10
The beam that is in thine own eye *Matt.* 7:3
If one be smitten against the other it shall be broken *Sir.* 13:2

Shakespeare
Let the world slide *Taming of the Shrew* Induction, 1
Hoops of steel *Hamlet I,* 3
No hinge, nor hoop to hang a doubt on *Othello* III, 3
The wheel has come full circle *King Lear* V, 3
Like a circle in the water . . . by broad spreading, it disperse to nought *King Henry VI, Part 1* I, 2

9   CIVIL ENGINEERING, MECHANICAL ENGINEERING AND MINING

Nursery Rhymes
Here is the church and here is the steeple
I'm the king of the castle
London Bridge is falling down
That lay in the house that Jack built
And the houses are built without walls
For want of a nail the shoe was lost

Proverbs and Sayings
A chip off the old block
A man's house is his castle
All roads lead to Rome
A bad workman blames his tools
As dead as a doornail
Every man to his trade
From pillar to post
Jack of all trades and master of none
Let the cobbler stick to his last
Necessity is the mother of invention
Rome was not built in a day
To build castles in the air
To carry coals to Newcastle
Walls have ears
Windmills in one's head

Biblical
The windows of heaven *Gen.* 7 : 11
Straw to make brick *Exod.* 5 : 7
The wheels of his chariots *Judg.* 5 : 28
Ye everlasting doors *Ps.* 24 : 7
Their houses shall continue forever *Ps.* 49 : 11
They have digged a pit before me *Ps.* 57 : 6
Went a whoring with their own inventions *Ps.* 106 : 39
The head stone of the corner *Ps.* 118 : 22
Within thy walls . . . within thy palaces *Ps.* 122 : 7
Build the house *Ps.* 127 : 1
Chains *Ps.* 149 : 5
Whoso diggeth a pit *Prov.* 26 : 27
They have sought out many inventions *Eccles.* 7 : 29
He that diggeth a pit *Eccles.* 10 : 8
The grinders cease *Eccles.* 12 : 3
The wheel broken at the cistern *Eccles.* 12 : 6
Them that join house to house *Isa.* 5 : 8
Crooked shall be made straight, and the rough places plain *Isa.* 40 : 4
As if a wheel had been in the midst of a wheel *Ezek.* 10 : 10
Wide is the gate, and broad is the way *Matt.* 7 : 13
The stone which the builders rejected *Matt.* 21 : 42
Go out into the highways and hedges *Luke* 14 : 23
The street which is called straight *Acts* 9 : 11
Not in tables of stone *2 Cor.* 3 : 3
An house not made with hands *2 Cor.* 5 : 1

Here we have no continuing city *Heb.* 13 : 14
The street of the city was pure gold *Rev.* 21 : 21

Shakespeare
More water glideth by the mill *Titus Andronicus* II, 1
A weathercock on a steeple *The Two Gentlemen of Verona* II, 1
House built on another man's ground *The Merry Wives of Windsor* II, 2
Laid on with a trowel *As You Like It* I, 2
Your only jig maker *Hamlet* III, 2
The engineer hoist with his own petard *Hamlet* III, 4
Delve one yard below their mines *Hamlet* III, 4
Mechanic slaves, with greasy aprons, rules and hammers *Antony and Cleopatra* V, 2
What the inside of a church is made of *King Henry IV, Part 1* III, 3

10   ZOOLOGY

Nursery Rhymes
Pussy's in the well
A frog he would a wooing go
The cat and the fiddle
The little dog laughed
The mouse ran up the clock
All the king's horses
Ladybird, ladybird
The lion and the unicorn
There came a big spider
What will poor Robin do then?
To get her poor dog a bone
Pussy cat, pussy cat, where have you been?
Upon a white horse
Four and twenty blackbirds
Down came a blackbird
Three blind mice
Frogs and snails and puppy dog's tails
Who killed Cock Robin?

Proverbs and Sayings
A bird in the hand is worth two in the bush
A cat has nine lives
A cat may look at a king

Barking dogs seldom bite
Birds of a feather flock together
Every dog has his day
He has a bee in his bonnet
A fish out of water
Kill two birds with one stone
Let sleeping dogs lie
Love me, love my dog
Never look a gift horse in the mouth
Never swap horses while crossing the stream
One can lead a horse to water
The early bird catches the worm
The last straw breaks the camel's back
Snake in the grass
To keep the wolf from the door
To stir up the hornet's nest
Like water off a duck's back
When the cat is away the mice will play
Like a cat on hot bricks

Biblical
Now the serpent *Gen.* 3 : 1
But the dove found no rest *Gen.* 8 : 9
Chastise you with scorpions *1 Kgs.* 12 : 11
The fowl of the air and the fish of the sea *Ps.* 8 : 8
I am a worm *Ps.* 22 : 6
Strong bulls of Bashan *Ps.* 22 : 12
A moth fretting a garment *Ps.* 39 : 12
Wings like a dove *Ps.* 55 : 6
Like a deaf adder *Ps.* 58 : 4
A portion for foxes *Ps.* 63 : 10
The horn of a unicorn *Ps.* 92 : 10
Pelican of the wilderness . . . owl of the desert . . . sparrow alone
upon the house top *Ps.* 102:6
The young lions roar *Ps.* 104 : 21
The strength of the horse *Ps.* 147 : 10
In the sight of the bird *Prov.* 1 : 17
Go to the ant *Prov.* 6 : 16
As the ox goeth to the slaughter *Prov.* 7 : 22
A whip for the horse, a bridle for the ass *Prov.* 26 : 3
Bold as a lion *Prov.* 28 : 1
An eagle in the air . . . a serpent upon the rock *Prov.* 30 : 19
The spider taketh hold *Prov.* 30 : 28

Dead flies *Eccles.* 10 : 1
The grasshopper shall be a burden *Eccles.* 12 : 5
The voice of the turtle is heard *S. of S.* 2 : 10
The wolf also shall dwell with the lamb *Isa.* 11 : 6
Habitation of dragons . . . owls *Isa.* 34 : 13
With wings as eagles *Isa.* 40 : 31
Or the leopard his spots *Jer.* 13 : 23
The palmerworm hath left hath the locust eaten *Joel* 1 : 4
O generation of vipers *Matt.* 3 : 7
The foxes have holes and the birds of the air have nests *Matt.* 8 : 20
Wise as serpents, and harmless as doves *Matt.* 10 : 16
Are not two sparrows sold for a farthing *Matt.* 10 : 29
It is easier for a camel *Matt.* 19 : 24
Which strain at a gnat, and swallow a camel *Matt.* 23 : 24
There will the eagles be gathered together *Matt.* 24 : 28
Where the worm dieth not *Mark* 9 : 44
As a roaring lion *1 Pet.* 5 : 8
Four beasts full of eyes before and behind *Rev.* 4 : 6
Behold, a pale horse *Rev.* 6 : 8

Shakespeare
The eagle suffers little birds to sing *Titus Andronicus* IV, 4
Bit with an envious worm *Romeo and Juliet* I, 1
Afflicted with these strange flies *Romeo and Juliet* II, 4
And laugh, like parrots *The Merchant of Venice* I, 1
A harmless necessary cat *The Merchant of Venice* IV, 1
The toad, ugly and venomous *As You Like It* II, 1
Quills upon the fretfull porcupine *Hamlet* I, 5
Like a crab *Hamlet* II, 2
Backed like a weasel *Hamlet* III, 2
Very like a whale *Hamlet* III, 2
The poor beetle that we tread upon *Measure for Measure* III, 1
A moth of peace *Othello* I, 3
The armed rhinoceros *Macbeth* III, 4
Mastiff, greyhound, mongrel grim, hound or spaniel *King Lear* III, 6
I marvel how the fishes live in the sea *Pericles* II, 1
As quarrelous as the weasel *Cymbeline* III, 4
To rouse a lion than to start a hare *King Henry IV, Part 1* I, 3
This fawning greyhound *King Henry IV, Part 1* I, 3
As merry as crickets *King Henry IV, Part 1* II, 4
Wrathful dove, or most magnanimous mouse *King Henry IV, Part 2* III, 2

11 AGRICULTURE

Nursery Rhymes
Baa, baa, black sheep
Nor yet feed the swine
Goosey, goosey gander
I had a little nut tree
Little Bo-Beep has lost her sheep
The sheep's in the meadow, the cow's in the corn
Mary had a little lamb
How does your garden grow?
Oranges and lemons
Peter Piper picked a peck of pickled pepper
A pocket full of rye
This is the farmer sowing his corn
This little pig went to market
'I'm going a milking, sir,' she said
Here we come gathering nuts in May

Proverbs and sayings
Grist to the mill
As well be hanged for a sheep as a lamb
Call a spade a spade
Do not count your chickens before they are hatched
To bolt the stable door after the horse has gone
Make hay while the sun shines
Never buy a pig in a poke
No rose without a thorn
Pigs might fly
To put the cart before the horse

Biblical
An olive leaf *Gen.* 8 : 11
A ram caught in the thicket *Gen.* 22 : 13
Seven fat kine *Gen.* 41 : 20
Corn in Egypt *Gen.* 42 : 1
A kid in his mother's milk *Exod.* 23 : 19
Muzzle the ox when he treadeth out the corn *Deut.* 25 : 4
The apple of his eye *Deut.* 32 : 10
Plowed with my heifer *Judg.* 14 : 18
As sheep that have not a shepherd *1 Kgs.* 22 : 17
Flourishing like a green bay tree *Ps.* 37 : 36
The cattle upon a thousand hills *Ps.* 50 : 10

Like grass which groweth up *Ps.* 90:5
The snare of the fowler *Ps.* 91:3
As grass: as a flower of the field *Ps.* 103:15
Sow in tears, shall reap in joy *Ps.* 126:3
A time to plant *Eccles.* 3:1
Shall not sow . . . shall not reap *Eccles.* 11:4
The almond tree shall flourish *Eccles.* 12:5
As the lily among thorns *S. of S.* 2:2
The flowers appear on the earth *S. of S.* 2:10
That spoil the vines *S. of S.* 2:15
An heap of wheat *S. of S.* 7:2
The ox knoweth his owner *Isa.* 1:3
Plowshares . . . pruning hooks *Isa.* 2:4
The grass withereth, the flower fadeth *Isa.* 40:7
Feed his flock like a shepherd *Isa.* 40:11
As a lamb to the slaughter *Isa.* 53:7
The harvest is past *Jer.* 8:20
Under his vine and under his fig-tree *Mic.* 4:4
Woe to the idle shepherd *Zech.* 11:17
The lilies of the field *Matt.* 6:28
Pearls before swine *Matt.* 7:6
In sheep's clothing, but inwardly they are ravening wolves *Matt.* 7:15
By their fruits ye shall know them *Matt.* 7:20
The harvest truly is plenteous *Matt.* 9:37
Some seeds fell by the wayside *Matt.* 13:4
A grain of mustard seed *Matt.* 17:20
Before the cock crow *Matt.* 26:34
I see men as trees, walking *Mark* 8:24
The husks that the swine did eat *Luke* 15:16
Bring hither the fatted calf *Luke* 15:23
I cannot dig *Luke* 16:3
A thorn in the flesh *2 Cor.* 12:7
Whatsoever a man soweth, that shall he also reap *Gal.* 6:7
Whose talk is of bullocks *Sir.* 38:25

Shakespeare
Small choice in rotten apples *Taming of the Shrew* I, 1
Do paint the meadows with delight *Love's Labour's Lost* V, 2
A rose, by any other name *Romeo and Juliet* II, 2
Two grains of wheat *The Merchant of Venice* I, 1
The savage bull doth bear the yoke *Much Ado About Nothing* I, 1
But keep a farm, and carters *Hamlet* II, 2

There's husbandry in heaven *Macbeth* II, 1
All my pretty chickens *Macbeth* IV, 3
Corn for the rich man only *Coriolanus* I, 1
The strawberry grows underneath the nettle *King Henry V* I, 1
A horse! a horse! *King Richard III* V, 4

12 TEXTILES

Nursery Rhymes
Sit by the fire and spin
But sit on a cushion and sew a fine seam

Proverbs and Sayings
A stich in time saves nine
As mad as a hatter
Cut your coat according to your cloth
If the cap fits, wear it
The coat makes the man
To look for a needle in a haystack
You cannot make a silk purse out of a sow's ear
To laugh up ones sleeve

Biblical
Sewed fig leaves together and made themselves aprons *Gen.* 3:7
A coat of many colours *Gen.* 37:3
This line of scarlet thread *Josh.* 2:18
Swifter than a weaver's shuttle *Job* 7:6
A moth fretting a garment *Ps.* 39:12
Neither do they spin *Matt.* 6:28
The hem of His garment *Matt.* 14:36
The eye of a needle *Matt.* 19:24
A wedding garment *Matt.* 22:11
Clothed in purple and fine linen *Luke* 16:19
Were white like wool *Rev.* 1:14

Shakespeare
Taffeta phrases, silken terms precise *Love's Labour's Lost* V, 2
That I were a glove *Romeo and Juliet* II, 2
As the fashion of his hat *Much Ado About Nothing* I, 1
Fashion wears out more apparel *Much Ado About Nothing* III, 3
Life is a shuttle *The Merry Wives of Windsor* V, 1
A king of shreds and patches *Hamlet* III, 4

The web of our life is of a mingled yarn *All's Well That Ends Well* IV, 3
I will wear my heart upon my sleeve *Othello* I, 1
Unpaid-for silk *Cymbeline* III, 3
Perfumed like a milliner *King Henry IV, Part 1* I, 3
But a shirt and a half *King Henry, IV Part 1* IV, 2

13   DIETETICS

Nursery Rhymes
I'll grind his bones to make my bread
And the dish ran away with the spoon
Hot cross buns
Jack Sprat could eat no fat
Some gave them white bread
Eating a Christmas pie
Eating her curds and whey
Sings for his supper
Pat-a-cake, pat-a-cake, baker's man
Pease porridge hot
She made some tarts
Simple Simon met a pieman
Baked in a pie
Eating bread and honey
Sugar and spice

Proverbs and Sayings
As like as two peas in a pod
Butter would not melt in his mouth
Do not put all your eggs into one basket
Enough is as good as a feast
Half a loaf is better than none
I know on which side my bread is buttered
It is no use crying over spilt milk
It takes the gilt off the gingerbread
One man's meat is another man's poison
Other fish to fry
Out of the frying pan into the fire
Promises are like pie-crust
Soft words butter no parsnips
The proof of the pudding is in the eating
To make two bites at a cherry

Too many cooks spoil the broth
What is sauce for the goose is sauce for the gander
You cannot have your cake and eat it
Your bread is buttered on both sides

Biblical
Shalt thou eat bread *Gen.* 3 : 19
Milk and honey *Exod.* 3 : 8
Butter in a lordly dish *Judg.* 5 : 25
Came forth meat . . . sweetness *Judg.* 14 : 14
There is death in the pot *2 Kgs.* 4 : 40
To boil like a pot *Job* 41 : 31
Honey and the honeycomb *Ps.* 19 : 10
Smoother than butter *Ps.* 55 : 21
Gall for my meat *Ps.* 69 : 21
Wine that maketh glad the heart *Ps.* 104 : 15
Bread eaten in secret is pleasant *Prov.* 9 : 17
A dinner of herbs *Prov.* 15 : 17
Wine is a mocker *Prov.* 20 : 1
Feed me with food convenient for me *Prov.* 30:8
Go eat, and to drink, and to be merry *Eccles.* 8 : 15
Comfort me with apples *S. of S.* 2 : 5
The fathers have eaten sour grapes *Ezek.* 18 : 2
His meat was locusts and wild honey *Matt.* 3 : 4
Our daily bread *Matt.* 6 : 11
New wine into old bottles *Matt.* 9 : 17
Let us eat and drink for tomorrow we die *1 Cor.* 15 : 32
Drink no longer water, but use a little wine *1 Tim.* 5 : 23

Shakespeare
Of a cut loaf *Titus Andronicus* II, 1
To suck the sweets *Taming of the Shrew* I, 1
Nourishment which is called supper *Love's Labour's Lost* I, 1
It is most sharp sauce *Romeo and Juliet* II, 4
As an egg is full of meat *Romeo and Juliet* III, 1
The world's mine oyster *The Merry Wives of Windsor* II, 2
As dry as the remainder biscuit *As You Like It* II, 7
Food of sweet and bitter fancy *As You Like It* IV, 3
'Twas caviare to the general *Hamlet* II, 2
I am a great eater of beef *Twelfth Night* I, 3
Ginger shall be hot i' the mouth *Twelfth Night* II, 3
The milk of human kindess *Macbeth* I, 5
My salad days *Antony and Cleopatra* I, 5

Like cold porridge *The Tempest* II, 1
A joint of mutton *King Henry IV, Part 2* V 1
Distressful bread *King Henry V* IV, 1
Small herbs have grace *King Richard III* II, 3

Index